Calcium-Based Materials

Calcium-based natural minerals are important for a wide range of applications. Though these materials are available in nature, researchers are working toward developing them in the laboratory.

Calcium-Based Materials: Processing, Characterization, and Applications introduces the possibility of designing these materials for particular applications.

- Introduces a variety of calcium-based materials and discusses synthesis, growth, and stability
- Provides in-depth coverage of calcium carbonate
- Discusses applications of calcium-based minerals in different fields
- Includes details on synchrotron X-ray tools for case minerals

This comprehensive text is aimed at researchers in materials science, engineering, and bioengineering.

Emerging Materials and Technologies

Series Editor: Boris I. Kharissov

The *Emerging Materials and Technologies* series is devoted to highlighting publications centered on emerging advanced materials and novel technologies. Attention is paid to those newly discovered or applied materials with potential to solve pressing societal problems and improve quality of life, corresponding to environmental protection, medicine, communications, energy, transportation, advanced manufacturing, and related areas.

The series takes into account that, under present strong demands for energy, material, and cost savings, as well as heavy contamination problems and worldwide pandemic conditions, the area of emerging materials and related scalable technologies is a highly interdisciplinary field, with the need for researchers, professionals, and academics across the spectrum of engineering and technological disciplines. The main objective of this book series is to attract more attention to these materials and technologies and invite conversation among the international R&D community.

Chemistry of Dehydrogenation Reactions and Its Applications
Edited by Syed Shahabuddin, Rama Gaur, and Nandini Mukherjee

Biosorbents: Diversity, Bioprocessing, and Applications
Edited by Pramod Kumar Mahish, Dakeshwar Kumar Verma, and Shailesh Kumar Jadhav

Principles and Applications of Nanotherapeutics
Imalka Munaweera and Piumika Yapa

Energy Materials: A Circular Economy Approach
Edited by Surinder Singh, Suresh Sundaramuthy, Alex Ibhadon, Faisal Khan, Sushil Kansal, and S.K. Mehta

Tribological Aspects of Additive Manufacturing
Edited by Rashi Tyagi, Ranvijay Kumar, and Nishant Ranjan

Emerging Materials and Technologies for Bone Repair and Regeneration
Edited by Ashok Kumar, Sneha Singh, and Prerna Singh

Mechanics of Auxetic Materials and Structures
Farzad Ebrahimi

Nanomaterials for Sustainable Hydrogen Production and Storage
Edited by Jude A. Okolie, Emmanuel I. Epelle, Alivia Mukherjee, and Alaa El Din Mahmoud

Calcium-Based Materials: Processing, Characterization, and Applications
Edited by S.S. Nanda, Jitendra Pal Singh, Sanjeev Gautam, and Dong Kee Yi

Advanced Synthesis and Medical Applications of Calcium Phosphates
Edited by S.S. Nanda, Jitendra Pal Singh, Sanjeev Gautam, and Dong Kee Yi

For more information about this series, please visit:
www.routledge.com/Emerging-Materials-and-Technologies/book-series/CRCEMT

Calcium-Based Materials

Processing, Characterization, and Applications

Edited by
S.S. Nanda
Jitendra Pal Singh
Sanjeev Gautam
Dong Kee Yi

CRC Press
Taylor & Francis Group
Boca Raton London New York

CRC Press is an imprint of the
Taylor & Francis Group, an **Informa** business

Designed cover image: © Sanjeev Gautam

First edition published 2024
by CRC Press
2385 NW Executive Center Drive, Suite 320, Boca Raton FL 33431

and by CRC Press
4 Park Square, Milton Park, Abingdon, Oxon, OX14 4RN

CRC Press is an imprint of Taylor & Francis Group, LLC

ISBN: 978-1-032-41955-8 (hbk)
ISBN: 978-1-032-41957-2 (pbk)
ISBN: 978-1-003-36059-9 (ebk)

DOI: 10.1201/9781003360599

Typeset in Nimbus Roman
by KnowledgeWorks Global Ltd.

Dedication

Dedicated to the boundless spirit of exploration,
symbolized by smart materials forging new paths in
self-healing bioengineering.
May our quest for knowledge and innovation lead us not
only to reshape our world,
but also to unravel the mysteries of life beyond Earth.

Contents

Preface

Calcium-based natural minerals play a pivotal role across a wide spectrum of applications. While these materials are naturally occurring, ongoing research is dedicated to their laboratory synthesize, enabling tailored designs for specific applications. This pursuit explores the potential for optimizing these materials' properties to suit particular purposes.

This book serves as a comprehensive guide to various calcium-based natural materials. It places specific emphasize on calcium-based oxides, carbonates, phosphates, and hydroxides. Within its pages, you will find in-depth discussions on the synthesis, growth, and stability of these materials under diverse processing conditions. The exploration extends to a thorough examination of their applications.

In essence, this book encompasses the following key aspects:

- A comprehensive exploration of the fundamental structure, chemistry, synthesis, and properties of both natural and synthetic calcium-based biomaterials.
- An insightful analysis of their current and potential future applications in biomedical engineering, medicine, and environmental contexts.

The compilation of knowledge within these chapters aims to provide researchers, practitioners, and enthusiasts in the field with an authoritative resource to deepen their understanding of calcium-based biomaterials and to inspire innovative advancements in their utilization.

Editors

S.S. Nanda, PhD, was born in Odisha, India, and earned a BPharm (2007) and MPharm (2009) in pharmacy at Biju Patnaik University of Technology, Odisha. In 2009, he became an Assistant Professor at Vikas College of Pharmaceutical Sciences, Suryapet, Telengana. In 2012, he moved to Gachon University and earned a PhD (2015) in bionanotechnology, working with Prof. Dong Kee Yi. He has authored and co-authored more than 50 peer-reviewed international journal articles and worked as an inventor for one patent. In 2015, he became an Assistant Professor at Myongji University. His research interests include tissue engineering and applications of materials in nanomedicine.

Jitendra Pal Singh, PhD, is a Ramanujan Fellow in the Department of Sciences (Physics), Manav Rachna University, Faridabad, Haryana, India. In 2010, he earned a PhD at Govind Ballabh Pant University of Agriculture and Technology, Pantnagar, Uttarakhand. He has worked at Pohang Accelerator Laboratory, Pohang, Republic of Korea, and the Korea Institute of Science and Technology, Seoul, Korea (2014–2022); Inter-University Accelerator Centre, New Delhi (2010–2011); Taiwan SPIN Research Centre, National Chung Cheng University, Taiwan (2011–2012); and Krishna Engineering College, Ghaziabad, India (2012–2014). His research interests are irradiation studies in nanoferrites, thin films, and magnetic multilayers. In May 2022, he joined the faculty of the Department of Sciences (Physics), Manav Rachna University, Faridabad, India. He is actively working on the synthesis of ferrite nanoparticles and thin films and determining the magnetic, optical, and dielectric responses of ferrites. He also studies the irradiation and implantation effects of ferrite thin films and nanoparticles. Dr. Singh has authored one book, *Ion Beam Induced Defects and Their Effects in Oxide Materials* (Springer, 2022), and edited five books: *Ferrite Nanostructured Magnetic Materials* (Elsevier, 2023); *Application of Ferrite Nanostructures* (Elsevier, 2023); *Oxide for Magnetic Applications* (Elsevier, 2023); *Defect Induced Magnetism in Oxide Semiconductors* (Elsevier, 2023); and *Sol-Gel Method: Recent Advances* (In Tech, 2023). He is a Guest Editor for several journals, including *Journal of Alloys and Compounds* and *RSC Advances*. He is also a Topic Editor for the journal *Magnetism (MDPI)* and the Founding Editor-in-Chief of the journal *Prabha Materials Science Letters*. He has authored and co-authored more than

150 peer-reviewed international journal articles related to ferrites, carbonates, X-ray absorption spectroscopy, and X-ray imaging.

Sanjeev Gautam, PhD, leads an independent research group, the Advanced Functional Materials Laboratory, at Dr. S.S. Bhatnagar University Institute of Chemical Engineering and Technology, Panjab University, Chandigarh, India. He has more than 25 years' experience with more than 161 international publications (h-index = 31 with 4000+ citations). He earned a PhD (2007) in condensed matter physics at the Centre of Advanced Study in Physics, Panjab University. He worked as a grid computing administrator in the CMS (LHC, Geneva) research project (MCSE, 2001–2007). As a beamline scientist at the Korea Institute of Science and Technology, South Korea (2007–2014), he was awarded a star post-doc. At Panjab University, as an Assistant Professor (2014–present), he has received international and national grants in nanotechnology, sustainable energy, food technology, catalysts, environmental safety, and administrative duties. Dr. Gautam has supervised seven PhD students and 32+ master's theses and promoted undergraduate research at Panjab University. He serves as an editorial board member for *Scientific Reports* (Nature Publications), *Materials Letters* and *Materials Letters: X* (Elsevier), and *Heliyon* (Cell).

Dong Kee Yi, PhD, earned a PhD in materials science and engineering in 2003 at the Gwangju Institute of Science and Technology (Korea). He went through his post-doc fellow seasons at Brown University and IBN at Singapore from 2003 to 2005. He worked as a Senior Scientist at the Samsung Advanced Institute of Technology from 2005 to 2007. From 2007 to 2013, he was on the faculty of the Department of Bionanotechnology, Gachon University (Korea). In 2013, he joined the faculty of the Department of Chemistry, Myongji University. He has edited one book, *Nanobiomaterials: Development and Applications* (CRC Press, Taylor & Francis, 2013). He has authored or co-authored more than 150 peer-reviewed international journal articles and worked as an inventor for 33 international patents. He serves on the editorial board for *ISRN Nanotechnology* and is a reviewer for the journals of leading scientific societies, including the American Chemical Society, the Royal Society of Chemistry, and the American Institute of Physics. He also works as a research/technology evaluator and on several advisory panels for the Chinese, Korean, and Romanian governments. He also is a consultant for industrial organizations in Korea.

1 Overview of Calcium-Based Materials

S.S. Nanda and Dong Kee Yi
Department of Chemistry, Myongji University
Yongin, South Korea

Nanomedicine and tissue engineering overtures have shown great promise in overcoming the main issue confronting orthopedic trauma, including acute infection risk and low burial reconstruction [1]. Nanoparticles have antibacterial properties and can help heal injured tissue [1]. Recently, nanomedicines have shown osseointegration property and stimulating bone process [2]. These characteristics lead toward essential components of orthopedic surgery [2]. Most of the aging population require orthopedic implants, and its estimation is about 600,000 per annum in the USA. Orthopedic implants help bones to grow and prevent infections [3].

Researchers describe nanomedicine for orthopedics as (i) using scaffold with nanomedicine to repair bone and cartilage defects, (ii) improving osseointegration and reducing biofilm preparation by designing implant surfaces, (iii) prolonging drug delivery systems with chemotherapeutic agents and antibiotics, (iv) delivering controlled drugs to combat infections, and (v) providing diagnostic applications for musculoskeletal infections and oncology [4].

Using gold nanoparticles for drug delivery through nanotechnology has shown promising results. Studies have shown the property of gold in effectively delivering iontophoresis to treat tendinopathy, or the disease and injury to the tendons [5, 6]. Gold nanoparticles and calcium materials together have special properties and uses in various areas. Calcium compounds can influence the structural and mechanical properties of materials. By incorporating them into gold nanoparticle systems, researchers could create materials with improved strength, durability, or other specific characteristics (Figure 1.1).

Selenium nanoparticles have drawn attention as a potential material in orthopedic applications for immobilization and suspension forms [8]. Selenium nanoparticles attributed low toxicity to human cells and served as an attractive antimicrobial agent [9]. Selenium is an essential trace element of human health. Incorporating selenium nanoparticles into calcium-based nutritional supplements could potentially enhance the bioavailability of selenium, leading to improved health benefits. Selenium nanoparticles have been explored for their potential in

DOI: 10.1201/9781003360599-1

1

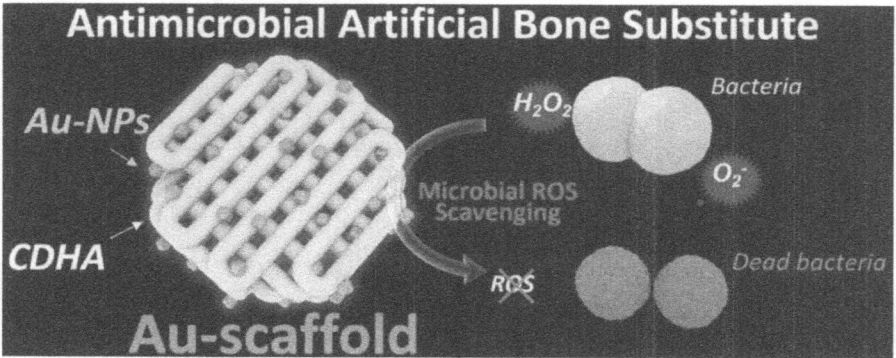

Figure 1.1 Immobilized scaffold with gold nanoparticles (Au-NPs) and ceramic (Calcium-deficient hydroxyapatite, CDHA) worked as a functional bone to antibacterial activity and scavenge microbial ROS at early stages. (Adapted from Reference [7] with permission.)

environmental remediation, such as the removal of heavy metals from water. Combining them with calcium materials might enhance their adsorption capacity and overall effectiveness.

Some examples of these new nanotherapies include a magnetite-hydroxyapatite composite, magnetite-enriched collagen hydroxyapatite biocompatible for bone grafting material, and a three-dimensional (3D) nanomagnetite-chitosan rod for local hypothermia that aids direct bonding to bones [10–12]. Nanotechnology for targeted drug therapy incorporated smaller particles (i.e., silver) into nanostructured materials, through the use of large particles (i.e., growth factors) into nanostructured materials [13–15]. Hydroxyapatite and tricalcium phosphate are extensively utilized as substitutes in bone grafting procedures to promote bone regeneration and support the seamless deposition of new bone tissue in the treatment of bone defects. It can make the entire artificial graft of HA or use HA as a surface coating. But there are high chances of rejection and donor incompatibility. To address these issues, researchers have explored the use of organic/inorganic composite materials that behave similarly to real bones. The combination of the inorganic phase along with the polymer component results in a better biomaterial. The inorganic phase provides strength and structural integrity to the material, allowing it to withstand mechanical stresses and maintain its shape. On the other hand, the polymer component adds toughness and flexibility, enabling the material to resist any deformation under impact or bending forces. Among all the forms of CaP, HA has the slowest degradation rate and excellent osteoconductive properties, and thus it is most preferred for bone grafts. Kim et al. [16] showed that a bone scaffold with up to 70% hydroxyapatite performs better. One approach to fabricating nano-hybrid bone tissue scaffolds involves in situ crystallization of HA within the structure of the polymer scaffold or by incorporating it into the scaffold via 3D printing techniques [17, 18]. Three-dimensional printing technology has also been utilized for the production of complex structures based on medical imaging, enabling the creation of personalized implants or grafts. In addition, these scaffolds can be used for localized drug delivery, which are then released in a controlled manner at the site of implantation or tissue regeneration. It is widely being used for visualization of damaged tissue. Martinez-Vazquez et al. [17]. formed porous silicon containing hydroxyapatite scaffolds using a 3D printing technique, for delivery and controlled release of vancomycin in short time.

1.1 CONCLUSION

Research has focused on developing osteoconductive materials for bone remodeling, as existing materials do not show the osteogenic properties required for bone regeneration. The mesenchymal stem cells derived from bone marrow play a crucial role in tissue repair due to their differentiation and regeneration abilities. Efforts are also being made to load bone scaffolds with drugs and use them for their controlled and effective release in damaged tissue without affecting the live tissue. Furthermore, the incorporation of magnetic materials with calcium phosphate nanoparticles into bone scaffolds has shown promise in promoting bone healing, particularly when combined with external stimulation (Figure 1.2). Overall, advancements in calcium-based biomaterials, including composite scaffolds and 3D printing technology, hold significant potential for orthopedic, dental, and oral surgery applications. These innovations in calcium-based biomaterials aim to improve the compatibility, functionality, and therapeutic effectiveness of bone grafts and implants.

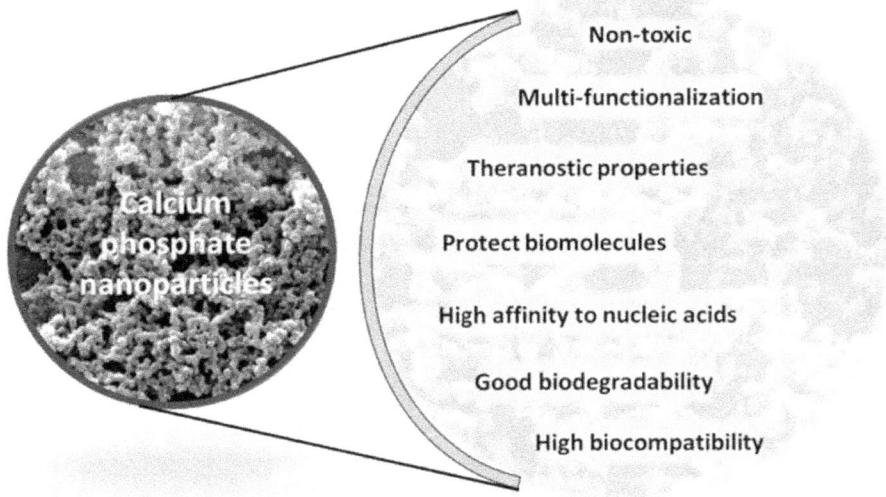

Figure 1.2 Benefits of calcium phosphate nanoparticles as a carrier system. (Adapted from Reference [19] with permission.)

REFERENCES

1. S. Behzadi, G. A. Luther, M. B. Harris, O. C. Farokhzad, M. Mahmoudi, *Biomaterials* 2017, **146**, 168–182.
2. M. Mazaheri, N. Eslahi, F. Ordikhani, E. Tamjid, A. Simchi, *International Journal of Nanomedicine* 2015, **10**, 6039–6054.
3. E. M. Christenson, K. S. Anseth, J. J. van den Beucken, C. K. Chan, B. Ercan, J. A. Jansen, S. Ramakrishna, *Journal of Orthopaedic Research* 2007, **25**, 11–22.
4. R. Garimella, A. E. Eltorai, *Journal of Orthopaedic* 2017, **14**, 30–33.
5. M. P. Sullivan, K. J. McHale, J. Parvizi, S. Mehta, *The Bone & Joint Journal* 2014, **96**, 569–573.
6. M. B. Dohnert, M. Venâncio, J. C. Possato, R. C. Zeferino, L. H. Dohnert, A. I. Zugno, T. F. Luciano, *International Journal of Nanomedicine* 2012, **7**, 1651–1657.
7. H. I. Kim, N. Raja, J. Kim, A. Sung, Y. J. Choi, H. S. Yun, H. Park, 2022, *Materials & Design*, **219**, 110793.
8. D. P. Biswas, N. M. O'Brien-Simpson, E. C. Reynolds, A. J. O'Connor, P. A. Tran, *Journal of Colloid and Interface Science* 2018, **515**, 78–91.
9. P. A. Tran, N. O'Brien-Simpson, J. A. Palmer, N. Bock, E.C. Reynolds, T. J. Webster, A. Deva, W. A. Morrison, A. J. O'Connor, *International Journal of Nanomedicine* 2019, **14**, 4613.
10. E. Andronescu, M. Ficai, G. Voicu, D. Ficai, M. Maganu, A. Ficai, *Journal of Materials Science: Materials in Medicine* 2010, **21**, 2237–2242.
11. Q. Hu, F. Chen, B. Li, J. Shen, *Materials Letters* 2006, **60**, 368–370.
12. S. Murakami, T. Hosono, B. Jeyadevan, M. Kamitakahara, K. Ioku, *Journal of the Ceramic Society of Japan* 2008, **116**, 950–954.
13. S. P. Arnoczky, O. Caballero, Y. N. Yeni, *JAAOS-Journal of the American Academy of Orthopaedic Surgeons* 2010, **18**, 445–448.
14. G. Wei, Q. Jin, , W. V. Giannobile, P. X. Ma, *Journal of Controlled Release* 2006, **112**, 103–110.
15. Z. C. Xing, W. P. Chae, M. W. Huh, L. S. Park, S. Y. Park, G. Kwak, I. K. Kang, *Journal of Nanoscience and Nanotechnology* 2011, **11**, 61–65.
16. H.-L. Kim, G.-Y. Jung, J.-H. Yoon, J.-S. Han, Y.-J. Park, D.-G. Kim, M. Zhang, D.-J. Kim, *Materials Science and Engineering: C* 2015, **54**, 20–25.
17. F. J. Martínez-Vázquez, M. V. Cabanãs, J. L. Paris, D. Lozano, M. Vallet-Regí, *Acta Biomaterialia* 2015, **15**, 200–209.
18. A. E. Jakus, R. N. Shah, *Journal of Biomedical Materials Research Part A* 2017, **105**, 274–283.
19. V. Sokolova, M. Epple, *Chemistry–A European Journal* 2021, **27**, 7471–7488

2 Calcium Phosphate
Synthesis and Applications

Sanjeev Gautam and Priyal Singhal
Advanced Functional Materials Laboratory, Dr. S.S. Bhatnagar
University Institute of Chemical Engineering and Technology
Panjab University, Chandigarh, India

2.1 INTRODUCTION

Biomaterials are man-made substances designed to function in intimate contact with biological systems and are utilized in therapy and diagnosis applications in treatment of diseased living tissues and organs. In an ideal scenario, a biomaterial intended to replace natural tissue should closely match its properties while ensuring the absence of any unwanted harmful effects within the living system.

Various calcium phosphate (CaP)-based synthetic biomaterials, like hydroxyapatite (HA), biphasic calcium phosphate (BCP), and tri-calcium phosphate (BCP) are extensively studied in the biomedical field. Trauma, osteoporosis, osteoarthritis, and surgical procedures are responsible for a significant number of musculoskeletal diseases, making it the second most prevalent cause of disabilities globally, as recognized by World Health Organization (WHO). Statistically, approximate 2.2 million individuals undergo bone grafting procedures annually for the treatment of bone regeneration and defects due to accidents, trauma or tumor resection but the absence of adequate bone replacement facilities has resulted in the loss of 1.2 million lives. Additionally, data suggests a global demand for dental implants in the range 10,000–30,000 annually [1,2].

Hard tissues, like bone and teeth, exhibit a high degree of mineralization. The mineral content in bone and dentin ranges from 45% to 70% by weight, while enamel reaches an impressive 96% by weight. The mineral phase of hard tissues is referred to as bioapatite which is a naturally occurring form of calcium phosphate found in biological systems. Its chemical formula $(Ca_5(PO_4)_3(OH))$ is similar to the hydroxyapatite $(Ca_{10}(PO_4)_6(OH)_2)$. The nature of the apatite phase found in bone mineral is nonstoichiometric. Bioapatite consists of various cations and anions like Sr^{2+}, Mg^{2+}, Na^+, K^+, Cl^-, F^-, CO_3^{2-}, etc. but it is OH^- deficient. These bioapatite crystals are made up of smaller units called crystallites. The size of these crystallites can vary among individuals depending on factors such as age and genetics. For ages

DOI: 10.1201/9781003360599-2

0–25, the individual crystal domain within the bioapatite mineral structure has a diameter of approximately 28 nm. The chemical structure of natural bone mineral, at the nanolevel, consists of inorganic calcium phosphate based hexagonal apatite-like structures and organic collagen bone matrix. These components combined provide the bone its strength and flexibility.

Its similarity with natural bone mineral has made it perfect for bone-tissue engineering applications [3]. It has many uses in toothpaste and pharmaceutical industry.

The second most studied bioceramic after HA is tricalcium phosphate(TCP). In nature, there are two stable polymorphs of tri-calcium phosphate called β and α-TCP. Out of these, β-TCP is more stable at ordinary temperature and exhibits a rhombohedral symmetry group. Due to high reactivity of α-TCP at high temperatures, β-TCP converts to α-TCP at around 1125 °C. However, the third polymorph, α'-TCP, is highly unstable and is of limited practical significance as it is only observed at temperatures exceeding 1430 °C and rapidly converts back to α-TCP upon being cooled below the temperature threshold [4].

Biphasic calcium phosphate(BCP), which is the combination of HA and β-TCP, is considered a better option for biomaterials. However, one major limitation of BCP, is its poor functional and mechanical properties. To address this drawback, extensive research has been conducted on the isomorphic substitution of various ions within the lattice structure of HA and TCP. This approach aims to overcome the limitations and enhance the properties of BCP.

Another form of calcium phosphate studied is tetracalcium phosphate (TTCP) which is found as the mineral "hilgenstockite" formed in the industrial slag. It has a monoclinic crystal structure and contains calcium, phosphate and oxide ions. It is represented by the chemical formula $Ca_4(PO_4)_2O$. This compound is not very rich in phosphorus. It is formed through a high temperature solid-state reaction at 1300 °C. It involves the combination of calcium oxide (CaO) and phosphorus pentoxide (P_2O_5) ions in an oxygen rich and high temperature environment. The unique feature of TTCP is its higher Ca/P ratio compared to hydroxyapatite(HA). It is commonly used to obtain composite coatings by plasma spray technique. TTCP reacts with acidic calcium phosphates like dicalcium phosphate anhydrous (DCPA, monetite) or dicalcium phosphate dihydrate (DCPD, brushite) and gets completely soluble simultaneously to form HA as the end product. Due to these properties, TTCP is used for bone repair and regeneration as it forms a crucial component in self-setting cements. Upon contact with the physiological environment, the cement undergoes a setting reaction that transforms it into a calcium phosphate based matrix, that gradually hardens and has similar properties with natural bone. TTCP is a metastable compound as its synthesis is done in a controlled environment with low humidity or in a dry atmosphere to maintain its stability. One way to achieve this is through fast quenching, which involves rapidly cooling the synthesized TTCP to room temperature after the calcination process. This prevents it from transforming into HA and lime. Another approach to prevent decomposition is by ensuring the absence of moisture during the synthesis process. Moisture can trigger the hydrolysis of TTCP, leading to the formation of HA and lime [5].

$$3Ca_4(PO_4)_2O + H_2O \rightarrow 2Ca_5(PO_4)_3OH + 2CaO \qquad (2.1)$$

The pursuit of artificial bone substitute materials has been a driving force behind substantial advancements in the field of biomaterials. Both biological and synthetic bone grafts can be used for bone grafting and hard bone tissue restoration and regeneration therapies. Biological bone graft materials are biocompatible materials derived from natural sources used to promote bone healing. They are broadly classified into three types:

- *Autografts*: The grafts used are extracted from the patient's own body, mostly from the iliac crest.
- *Allografts*: The bone scaffold is made using hard tissues of a donor, generally obtained from cadavers.
- *Xenografts*: The grafts are derived from different species, mainly animals.

While biological materials closely mimic the properties of natural bone, there is always a high risk of immunorejection and microbial contamination. Also, there is a very limited supply of biological materials.

Synthetic bone graft materials are artificially created materials that are designed to mimic the properties of natural bone and promote bone regeneration. These synthetic implant materials are of three types (Figure 2.1).

First-generation materials, such as titanium or stainless steel, are bioinert, that is, they generate limited response to the presence of foreign implants while second generation materials like calcium phosphate based materials, form covalent or other types of chemical bonds with surrounding living tissues and are bioactive. In contrast, third generation materials interact with cells at a highly intricate level, influencing their behavior and functions at a molecular scale. They can actively promote desired cellular responses such as proliferation, differentiation, migration, and tissue regeneration.

2.2   OVERVIEW: CALCIUM

2.2.1   STRUCTURE

Calcium phosphate refers to a group of compounds composed of calcium ions (Ca^{2+}) and phosphate ions (PO_4^{3-}). The crystal structure differs as the stochiometric ratio of calcium phosphate changes.

HA exists in two phases- hexagonal and monoclinic. Hexagonal phase is preferred more as it has more resemblance with natural bone mineral. It has hexagonal shaped nano-crystallite structure. Researchers are aiming to achieve a material that closely mimics the characteristics of natural bone tissue by synthesizing heavily ion-doped HA as Hydroxyapatite is capable of preserving its crystal symmetry when the calcium-to-phosphorus (Ca/P) molar ratio remains at 1.3, compared to the ideal Ca/P ratio of 1.67. HA crystal can be doped with multiple ions and elements by replacing

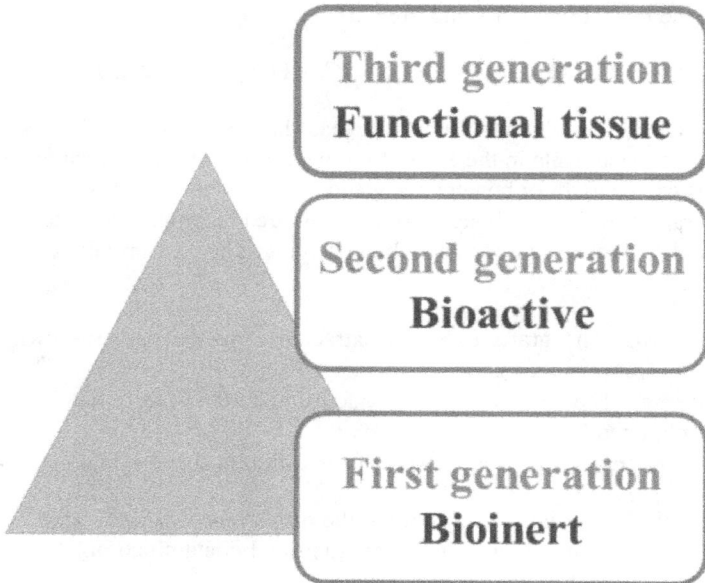

Third generation
Functional tissue

Second generation
Bioactive

First generation
Bioinert

Figure 2.1 Different generations of synthetic implant materials [3]. (Adapted with permission from Wu *et al. J. Hazardous Mater.* 387, 2020. Copyright 2020 Elsevier.)

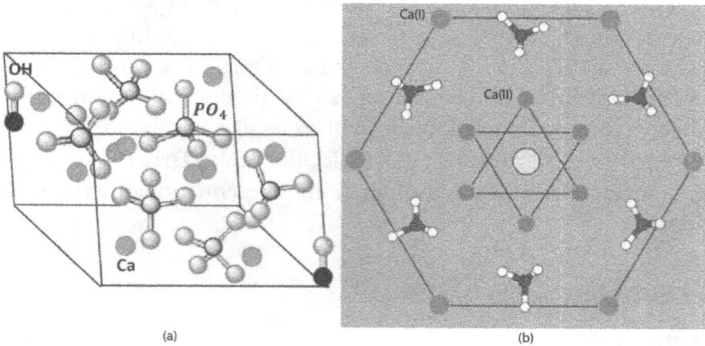

Figure 2.2 (a) Crystal structure of hydroxyapatite. (b) Typical diagram of unit cell of hydroxyapatite with hexagonal phase [6,7].

calcium ions with cations, and phosphate or hydroxyl ions by anions. The hexagonal HA exhibits *P63/m* symmetry (Figure 2.2) with specific locations for two different Ca atoms while monoclinic HA possesses *P21/b* crystallographic symmetry [6]. Ca(I) atoms have simple cubic structure and are positioned only at the edges, while Ca(II) atoms form an equilateral triangle with the ion channel at the center of the unit

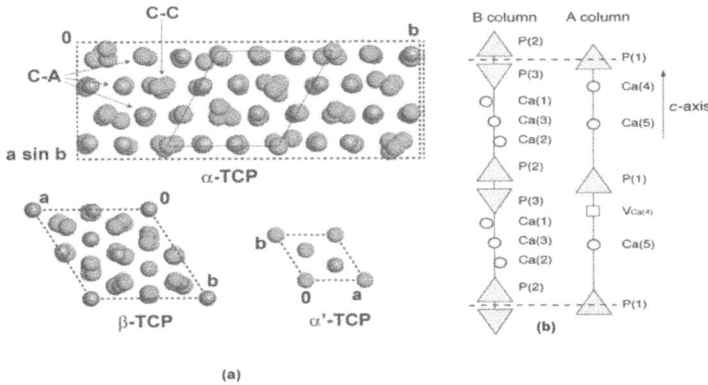

Figure 2.3 (a) Crystal structure of α, β, α'-tri calcium phosphate. (b) Expanded lattice arrangement of β-tricalcium phosphate [3,4].

cell. Among the components of HA, phosphate ions have a significant impact on the structure of the unit cell [8,9].

β-TCP consists of a 3D network of calcium and phosphate ions arranged in a rhombohedral unit cell. It belongs to the point group *R3c*. Its parameters are $a = 10.3961$ and $c = 37.3756$ Å. The lattice of β-TCP is arranged into two different columns, labeled as "A" and "B", running parallel to the "c" axis as shown in Figure 2.3. Within the structure, there are five non-equivalent calcium (Ca) sites. Each "A" column is coordinated to six "B" columns, forming an octahedral environment while each "B" column forms tetrahedral arrangement by bonding with four "A" columns. Both "A" and "B" columns are arranged in alternating layers [10].

In contrast, α- and α'-TCP have different crystal structures. In its monoclinic form, α-TCP exhibits a specific arrangement in its crystal lattice with the *P21/a* point group, and its lattice parameters are $a = 12.8592$, $b = 27.3542$, and $c = 15.2223$ Å. On the other hand, α'-TCP has a hexagonal point group of *P63/mmc*, and its lattice parameters are $a = 5.3507$ and $c = 7.6841$ Å (Figure 2.3).

In α-TCP, there are two types of columns: "C-C" and "C-A". The "C-C" column solely consists of calcium cations, while the "C-A" column contains both calcium cations and phosphate anions. One "C-C" column is surrounded by six "C-A" columns, while one "C-A" column is coordinated to six "C-C" and "C-A" columns. Both these columns are arranged in alternating layers. A notable distinction between α- and β-TCP lies in their structural composition. In β-TCP, there are no "C-C" columns but it has two types of "C-A" columns. α'-TCP exhibits a similar arrangement to α-TCP, with alternating "C-C" and "C-A" columns [11]. Table 2.1 represents the structural difference between different forms of TCP.

Monoclinic tetracalcium phosphate has *P21* space group symmetry with lattice parameters $a = 7.023$ Å, $b = 11.986$ Å, $c = 9.473$ Å and $b = 90.90°$ at 25°C. The structure of TTCP is very similar to HA and is described as pseudo-orthorhombic [14].

Table 2.1

Structural Data of Different forms of Tri-calcium Phosphate

Property	β-TCP	α-TCP	α'-TCP
Symmetry	Rhombohedral	Monoclinic	Hexagonal
Space group	R3C	P21/a	P63/mmc
a (nm)	1.04352	1.2859	0.53507
b (nm)	1.04352	2.7354	0.53507
c (nm)	3.74029	1.5222	0.7684
α	90	90	90
β	90	126.35	90
γ	120	90	120

Source: References [4, 12, 13].

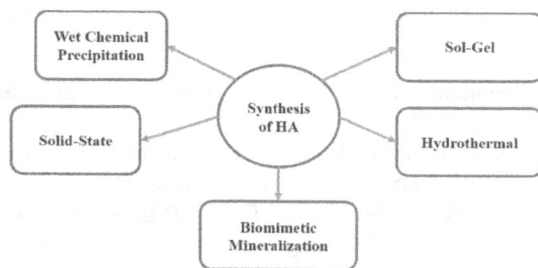

Figure 2.4   Synthesis methods of hydroxyapatite.

2.3   SYNTHESIS METHODS

The choice of method depends on factors such as desired particle characteristics, scalability, and intended applications. Figure 2.4 depicts some of the techniques used to obtain hydroxyapatite.

2.3.1   WET CHEMICAL PRECIPITATION

This method involves the solidification of HA from an aqueous solvent containing calcium and phosphate ions. Typically, a calcium precursor (such as calcium nitrate) and a phosphate precursor (such as diammonium hydrogen phosphate) are mixed together. The reaction leads to the formation of HA particles, which can be further processed and collected [15]. The formation of HA through precipitation is facilitated by controlling factors such as the concentrations of calcium and phosphate ions, pH level, temperature, and reaction time.

2.3.2   SOL-GEL METHOD

This method allows for the control of particle size, morphology, and composition of the synthesized HA. In this method, two different chemical reagents, namely

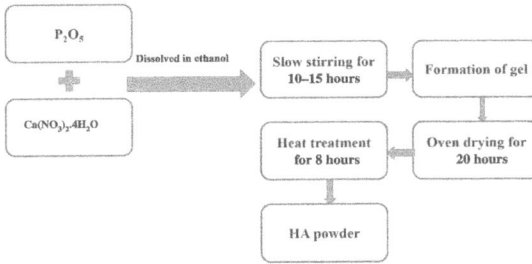

Figure 2.5   Schematic diagram showing sol–gel method to obtain hydroxyapatite.

phosphoric pentoxide (P_2O_5) and calcium nitrate tetrahydrate ($Ca(NO_3)_2 \cdot 4H_2O$), were utilized. To prepare the desired solution with a calcium-to-phosphorus (Ca/P) molar ratio of 1.67, first, absolute ethanol was used to dissolve P_2O_5 until a concentration of 0.5 mol/l was achieved. Second, another solution with a concentration of 1.67 mol/l was prepared by mixing $Ca(NO_3)_2 \cdot 4H_2O$ with ethanol. To ensure thorough mixing, the combined solution was stirred slowly for a period of 10 to 15 hours. As time passed, a transformation occurred, and the solution gradually formed a gel-like substance. The gel was subsequently subjected to a drying process in an electric oven at 80 °C in ambient air for 20 hours, then it undergoes heating for 8 hours which raises the temperature from 400 to 750 °C [16]. The heat treatment process occurs in two stages. In the first stage of the heat treatment, the temperature is gradually raised from 400 to 750 °C.

Synthesis via sol-gel occurs in three steps:

- Creating a 3D network between hydrolysed metal hydroxide/oxide species in an organic solvent.
- Formation of gel through condensation of solution followed by subsequent aging and drying to remove the excess liquid.
- Calcination of the dried gel [17].

This slow and controlled increase in temperature allows for the removal of any remaining moisture and volatile components from the gel. The second stage of the heat treatment involves maintaining the temperature at 750 °C for a certain period of time which allows for further consolidation and crystallization of the material. The complete process is shown in Figure 2.5.

2.3.3 HYDROTHERMAL METHOD

The hydrothermal method enables the synthesis of HA with controlled crystallinity and size. Hydrothermal synthesis involves the reaction of precursor materials in a high-pressure, high-temperature aqueous environment. Under these conditions, HA can be formed through mineralization processes. A mixture of calcite (0.80 g) and DCPA (1.64 g) is uniformly blended with 30 ml of deionized water. The purpose of this step is to ensure that the calcite and DCPA are evenly distributed throughout the mixture. To achieve this, a teflon container is used as the blending vessel. Once the mixture is prepared, the teflon container is sealed and placed inside an autoclave. The autoclave serves as a controlled environment where the mixture will undergo heat treatment. The temperature inside the autoclave is raised to specific temperatures ranging from 120 to 180 °C, and the mixture is exposed to these temperatures for different durations of time. This step is crucial for promoting chemical reactions and transformations within the mixture. After the designated treatment times, the resulting powder was collected from the teflon container. To remove any impurities or residual substances, the collected powder is washed with deionized water. Finally, the washed powder is dried at a temperature of 80 °C. This drying process removes any remaining moisture and prepares the powder for further analysis and characterization. [18]. The hydrothermal method involves the reaction of reagents at high temperature and pressure [19]. However, a significant drawback of this method is the difficulty in controlling particle size and morphology. To overcome these limitations, organic modifiers are employed. Organic modifiers improve the stability, reactivity and functionality of the chemical processes involved. They are of two types – chelating agents and surface-active agents or surfactants. Among them, ethylene-diamine-tetra-acetic acid (EDTA) is an organic surfactant widely used to stabilize emulsion in the hydrothermal technique [20]. By incorporating organic modifiers, it becomes possible to enhance control over particle size and morphology during the hydrothermal synthesis process.

2.3.4 SOLID STATE SYNTHESIS

In this synthesis method, calcium and phosphate powders are combined together. Then, the resulting mixture undergoes heat treatment at very high temperatures, typically above 800 °C, to promote the reaction between the precursors and the formation of HA. This method is relatively simple and cost-effective, but it requires high temperatures and may result in larger particle sizes.

Figure 2.6 shows different synthesis methods for formation of doped BCP. The solid state method involves mixing HA and β-TCP materials in the desired ratios.

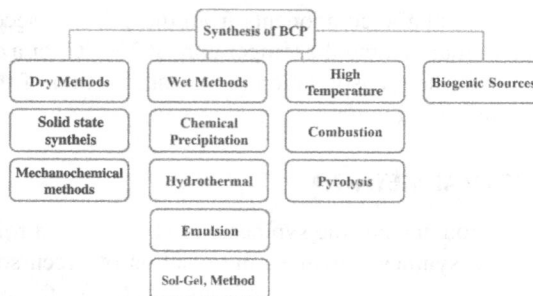

Figure 2.6 Synthesis methods to obtain doped biphasic calcium phosphate [3].

The powders can be synthesized separately. The HA and β-TCP compounds are then thoroughly combined to achieve homogeneous distribution using mechanical mixing techniques, such as ball milling or grinding. After the mechanical mixing step, the resulting mixture can be further processed as needed, such as by compaction or sintering, which involves heating the compacted mixture to high temperatures below the melting point of the powders (above 1000 °C) to facilitate solid-state reactions that transform mixed HA and β-TCP powders into a solid material, leading to the formation of BCP. The mechanochemical approach offers certain advantages over the solid-state reaction method, such as improved reproducibility and its suitability for large-scale production [21].

2.3.5 BIOMIMETIC MINERALIZATION

Biomimetic approaches aim to mimic the natural mineralization process of HA in the body. This method involves the nucleation and growth of HA on a substrate or template surface in a simulated physiological environment. Biomimetic mineralization can be achieved by controlling the solution conditions, such as pH, temperature, and ion concentrations, to promote the formation of HA with similar properties to natural bone mineral.

Two commonly used methods for synthesis of TCP are wet chemical precipitation method and solid state reaction method.

In chemical precipitation synthesis, typically, calcium hydroxide ($Ca(OH)_2$) and phosphoric acid (H_3PO_4) or dibasic ammonium phosphate (($NH_4)_2HPO_4$) are used as the calcium and phosphate sources, respectively. The precursors are dissolved in water to create a solution under controlled pH. By carefully controlling the pH, specific conditions are created that encourage the desired chemical transformation of the precursors into TCP. Once the TCP nanoparticles have formed, they are separated from the solution using a filtration process. This helps to isolate the solid particles from the liquid phase. The collected nanoparticles are then washed to remove any impurities or residual substances that may be present.

After the washing step, the TCP nanoparticles are dried. This involves removing the remaining water content from the particles, resulting in the formation of TCP precipitates in solid powder form [22]. The solid-state reaction method involves the direct reaction between calcium carbonate ($CaCO_3$) and dicalcium phosphate (DCP) or monocalcium phosphate (MCP) in a high-temperature furnace. The precursors are thoroughly mixed in the desired stoichiometric ratio and then undergo heat treatments at very high temperatures, typically between 1100 and 1400 °C for several hours. This process promotes the chemical reaction between the precursors, resulting in the formation of TCP. The obtained TCP product is subsequently cooled, grounded, and sieved to obtain the desired particle size [23].

2.3.6 EMULSION METHODS

The emulsion process carried out for synthesizing BCP uses two immiscible phases of calcium and phosphorus sources dispersed in each other. Generally, a multiple emulsion (W/O/W) is used to obtain HA powders. There are three types of emulsion:

- *Oil-in-water phase*: Oil droplets dispersed in continuous water phase.
- *Water-in-oil phase*: Water droplets dispersed in continuous oil phase.
- *Multiple emulsion*: Also called emulsion within an emulsion. It can be either oil-in-water-in-oil or water-in-oil-in-water.

Lim et al. [24] obtained nanocrystalline HA powders via emulsion process.

2.3.7   HIGH TEMPERATURE METHODS

Some elevated high temperature methods like pyrolysis and combustion are employed to partially burn the precursors. Pyrolysis is capable of producing highly crystalline HA [25], but it is associated with the drawback of secondary agglomeration formation. However, this can be mitigated by incorporating specific salts into the precursor solution [26]. In the combustion method, the reactants capable of undergoing exothermic redox reaction on their own are chosen which undergo combustion, in the presence of oxygen, resulting in the release of energy. To facilitate this reaction, an appropriate organic fuel is required. $Ca(NO_3)_2$ is commonly used as a precursor, along with fuels such as urea, citric acid, glycine, sucrose, etc., are employed for the synthesis of BCP [27, 28]. Ghosh et al. prepared HA nanocrystals through combustion method [29].

Biogenic sources can also be used to synthesize HA. Such products are more bioactive than synthetic BCP and closely resemble the properties of bioapatite.

Conventional methods for synthesizing TTCP involve mixing calcium carbonate ($CaCO_3$) and dicalcium phosphate anhydrate ($CaHPO_4$) in a specific ratio of calcium to phosphorus(Ca/P) of 2 [30]. Once the mixture is prepared, it is subjected to high-temperature conditions ranging from 1450 to 1500 °C for 6–12 hours. The high temperatures promote the solid state reaction between the calcium carbonate and dicalcium phosphate anhydrate, initiating a chemical transformation. This reaction leads to the formation of TTCP as the main product. The solid-state reaction can be represented by the following equation:

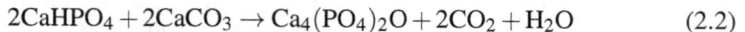

$$2CaHPO_4 + 2CaCO_3 \rightarrow Ca_4(PO_4)_2O + 2CO_2 + H_2O \qquad (2.2)$$

To prevent the decomposition of TTCP into other secondary phases like HA, calcium oxide, $CaCO_3$, and β-tricalcium phosphate and to obtain highly pure TTCP, the mixtures must undergo fast quenching process after heating. By adjusting the process temperatures, it has been observed that at lower temperatures, undesired products like HA or TCP appear. However, as the process temperature is increased, TTCP becomes the dominant product. By subjecting the precursors to the following reaction at a temperature of 1350 °C, TTCP product with minimal impurities and well defined crystalline structure is obtained. TTCP can be obtained at a much lower temperature of 1200 °C through the following reaction:

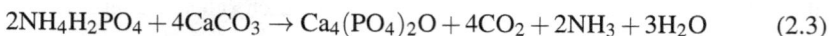

$$2NH_4H_2PO_4 + 4CaCO_3 \rightarrow Ca_4(PO_4)_2O + 4CO_2 + 2NH_3 + 3H_2O \qquad (2.3)$$

However, the purity of the product obtained at 1200 °C was lower compared to that obtained at 1350 °C.

2.4 CHARACTERIZATION

Various characterization techniques like XRD, TEM, SEM, FT-IR are used to study about the structure, composition and morphology of calcium phosphate. Figures 2.7 and 2.8 [31] show SEM and FTIR analysis of synthesized hydroxyapatite at various sintering temperatures (900, 1000, 1100 °C), respectively.

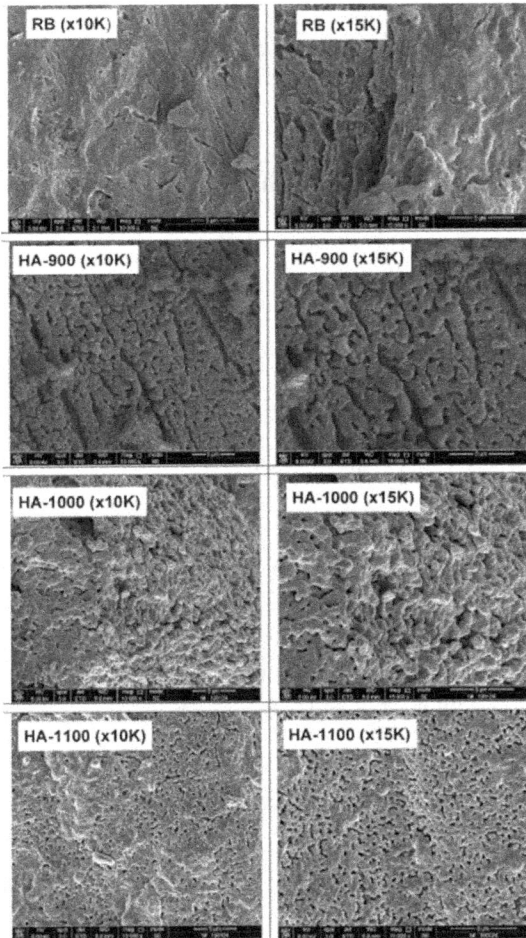

Figure 2.7 SEM micrographs of hydroxyapatite at diff temp and magnification [31]. (Adapted with permission from Abifarin J. K., et al. "Experimental data on the characterization of hydroxyapatite synthesized from biowastes." Data in brief 26 (2019): 104485. Copyright 2019 Elsevier.)

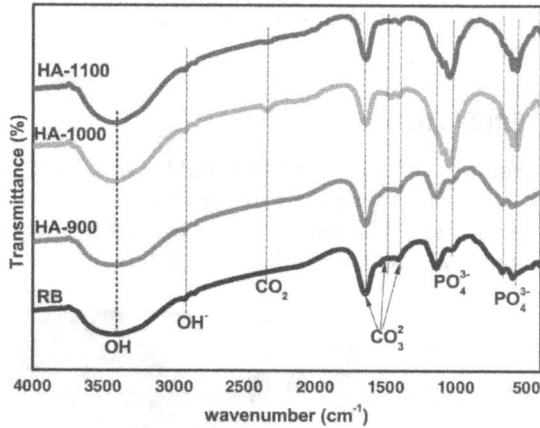

Figure 2.8 FTIR spectrograph of hydroxyapatite [31]. (Adapted with permission from Abifarin J. K., et al. "Experimental data on the characterization of hydroxyapatite synthesized from biowastes." Data in brief 26 (2019): 104485. Copyright 2019 Elsevier.)

Table 2.2

Wavenumbers(cm^{-1}), Associated Chemical Group and Explanation of FTIR Spectra of Hydroxyapatite [31,32]

HA-900 (cm^{-1})	HA-1000 (cm^{-1})	HA-1100 (cm^{-1})	Chemical Group	Description
570,601	570,601	570,601	PO_4^{3-}	Bending mode
632,3417	632,3417	632,3417	OH^-	Proves presence of HA
1,049,1095	1,049,1095	1,049,1095	PO_4^{3-}	Antisymmetric stretching mode
1,411,1465	1,411,1465	1,411,1465	CO_3^{2-}	Due to $CaCO_3$ component of HA
2345	2345	2345	CO_2	Released during heat treatment

SEM images show that the microstructure of hydroxyapatite becomes progressively denser with higher sintering temperatures. As the temperature during sintering increases, the particles begin to fuse and bond together more tightly. This fusion process leads to a reduction in the number of void spaces or pores between the particles. Consequently, the overall density of the hydroxyapatite structure increases.

When the SEM images are analyzed, it becomes apparent that at higher sintering temperatures, the hydroxyapatite particles exhibit a more compact arrangement. The boundaries between individual particles become less distinguishable as they merge together, resulting in a denser and more uniform microstructure.

The characteristic bands associated with the $CaCO_3$ component are typically observed at specific wavenumbers, around 1411 and 1465 cm^{-1}. These bands indicate the presence of carbonate ions within the hydroxyapatite structure. During the sintering process, additional changes in the vibrational bands around 2345 and 2353 cm^{-1} can be observed (Table 2.2).

This suggests that the carbonate groups within the hydroxyapatite structure undergo changes and release CO_2 as a byproduct during the sintering process. Another significant feature in the infrared spectrum is a distinct broad band observed in the range of $1000–1100$ cm^{-1}. This band corresponds to the asymmetric stretching vibrational mode due to the phosphate group, is an indicator of the presence of hydroxyapatite. Additionally, a distinctive band for all samples is observed within the range of $570–565$ cm^{-1}. This particular band represents the symmetric stretching vibration of the phosphorous–oxygen (P–O) bonds present in the phosphate (PO_4) group.

Figure 2.9 shows results from different characterization techniques for tri-calcium phosphate polymorphs. X-ray diffraction helps to differentiate between different forms of TCP. By analyzing the XRD pattern, the presence of specific peaks and their intensities can be used to identify different crystal phases of tricalcium phosphate.

XANES (X-ray absorption near-edge structure) spectroscopy is a technique used to study the local electronic and geometric structures of tricalcium phosphate. It provides valuable insights into the chemical bonding and coordination of atoms within the material.

Raman spectroscopy provides information on its molecular vibrations and crystal symmetry.

Raman spectra can reveal characteristic peaks that correspond to specific molecular bonds and help identify different forms or impurities in tricalcium phosphate samples.

FTIR (Fourier-transform infrared) spectroscopy is used to study the vibrational modes of tricalcium phosphate. By measuring the absorption of infrared light, FTIR spectra can identify functional groups present in the material, such as phosphate groups (PO_4) and hydroxyl groups (OH).

Figures 2.10 and 2.11 show XRD and TEM analysis of synthesized TTCP powder with ultrafine nanoparticles.

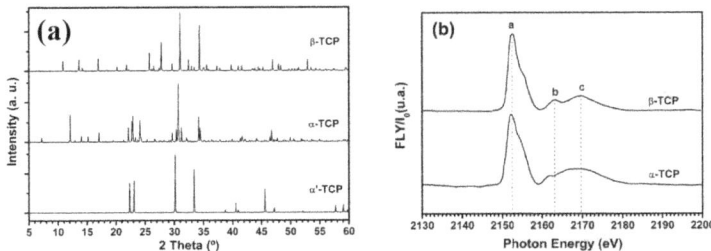

Figure 2.9 (a) XRD. (b) XANES spectra of polymorphs of tri-calcium phosphate [33]. (Adapted with permission from (a) Carrodeguas Ral G., and Salvador De Aza. "Tricalcium phosphate: Synthesis, properties and biomedical applications." Acta biomaterialia 7.10 (2011): 3536–3546. Copyright 2011 Elsevier. (b) Demirkiran Hande, et al. "XANES analysis of calcium and sodium phosphates and silicates and hydroxyapatite Bioglass 45S5 co-sintered bioceramics." Materials Science and Engineering: C 31.2 (2011): 134–143. Copyright 2011 Elsevier.)

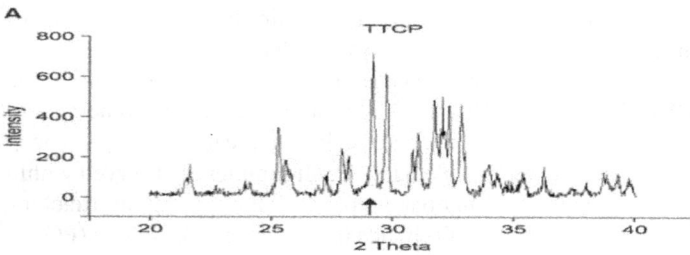

Figure 2.10 XRD analysis of tetra-calcium phosphate powder [34]. (Adapted with permission from Kwon Ki-Dae, et al. "The Effect of cefazolin on mechanical properties and antibacterial reactions of calcium phosphate cement." Journal of the Korean Orthopaedic Association 46.4 (2011): 273–281. Copyright 2011.)

Figure 2.11 TEM of tetra-calcium phosphate powder [35]. (Adapted with permission from Qin Tian, et al. "Bioactive tetracalcium phosphate scaffolds fabricated by selective laser sintering for bone regeneration applications." Materials 13.10 (2020): 2268. Copyright 2020.)

TEM (Transmission Electron Microscopy) analysis of tetracalcium phosp (Figure 2.11) involves the use of a high-resolution electron microscope to examine the material at the nanoscale. TEM allows for the visualization of the internal structure, morphology, and size distribution of tetracalcium phosphate particles. By capturing electron images, TEM provides detailed information about the arrangement of atoms and crystal defects within the material, allowing for a more precise characterization of its nanostructure.

2.5 BIOMEDICAL APPLICATIONS OF CALCIUM PHOSPHATE

Calcium phosphate is a versatile material that finds numerous applications in the field of medicine. This is primarily attributed to its exceptional biocompatibility, osteoconductivity, and close resemblance to the mineral composition of natural bone. So, it finds many clinical applications in orthopedic and dental implants and is also used in the drug delivery systems.

2.5.1 ARTIFICIAL BONE GRAFTS

Hydroxyapatite and tricalcium phosphate are extensively utilized as substitutes in bone grafting procedures to promote bone regeneration and support the seamless deposition of new bone tissue in the treatment of bone defects. The artificial graft can be made entirely of HA or use HA as a surface coating. But there are high chances of rejection and donor incompatibility. To address these issues, researchers have explored the use of organic/inorganic composite materials that behave similar to real bones. The combination of the inorganic phase along with the polymer component results in a better biomaterial. The inorganic phase provides strength and structural integrity to the material, allowing it to withstand mechanical stresses and maintain its shape. On the other hand, the polymer component adds toughness and flexibility, enabling the material to resist any deformation under impact or bending forces. Among all the forms of CaP, HA has the slowest degradation rate and excellent osteoconductive properties, thus it is most preferred for bone grafts. Kim et al. [36] showed that a bone scaffold with upto 70% hydroxyapatite performs better. One approach to fabricating nano-hybrid bone tissue scaffolds involves *in situ* crystallization of HA within the structure of the polymer scaffold or by incorporating it into the scaffold via 3D printing techniques [37, 38].

Three-dimensional printing technology has also been utilized for the production of complex structures based on medical imaging, enabling the creation of personalized implants or grafts. In addition, these scaffolds can be used for localized drug delivery which are then released in a controlled manner at the site of implantation or tissue regeneration. It is widely being used for visualization of damaged tissue. Martinez-Vazquez et al. [37] formed porous Silicon contained hydroxyapatite scaffolds using a 3D printing technique, for delivery and controlled release of vancomycin in short time.

Research has focused on developing osteoconductive materials for bone remodeling, as many existing materials do not show osteogenic properties required for bone regeneration. The mesenchymal stem cells derived from bone marrow play a crucial role in tissue repair due to their differentiation and regeneration abilities. Efforts are also being made to load bone scaffolds with drugs and use them for their controlled and effective release in damaged tissue without affecting the live tissue. Furthermore, the incorporation of magnetic materials into bone scaffolds has shown promise in promoting bone healing, particularly when combined with external stimulation.

Overall, advancements in biomaterials, including composite scaffolds and 3D printing technology, hold significant potential for orthopedic, dental, and oral surgery applications. These innovations aim to improve the compatibility, functionality, and therapeutic effectiveness of bone grafts and implants.

2.5.2 DENTAL IMPLANTS COATINGS

Calcium phosphate is used as a coating material on dental implant surfaces to improve their osseointegration. These coatings enhance the adhesion and proliferation of bone cells. When bone cells adhere and grow on the implant surface, they form a strong bond, creating a secure foundation between the implant surface and surrounding bone tissue. Also, the proliferation of bone cells encourages their growth and multiplication. This cellular activity along with osseointegration supports the formation of new and strong bone tissue around the implant, which further contributes to its stability and longevity.

Metal implants are usually not biocompatible, therefore hydroxyapatite (HA) coatings are extensively used on metal implants to improve its biocompatibility and osteoconductive properties [39]. Various techniques like plasma spraying, sol-gel deposition and sputtering have been used to prepare HA coatings. Below are some characteristics of a good HA coating-

- Higher crystalline behavior
- Strong adhesion to the substrate
- Sufficient porosity

Plasma spraying is a commonly used and cost-effective method for coating implant surfaces but this technique has limited practical applications in biomedical field. Plasma-sprayed coatings show structural and chemical inconsistencies. The coating does not have strong adhesion with the metal implant surface and microcracks can also be seen on its surface [40].

The sol–gel technique, on the other hand, is a simple and cost-efficient method for preparing thin films. It allows for the production of homogeneous films with a fine grain structure at low temperatures.

Recently, carbon nanotubes are being used to obtain crack-free and homogeneous HA coatings with higher crystallinity and biocompatibility. In addition to biocompatibility, there is a growing demand for implant surfaces with antibacterial properties to avoid risk of potential infection after implantation. The infection mainly occurs as the bacteria surrounds the implant surface and forms a film, thus damaging the implant site with sepsis. Carbon nanotubes are being explored for their antibacterial properties, so it can be used for localized delivery of antibiotics.

2.5.3 DRUG DELIVERY

Calcium phosphate based porous nanoparticles and microparticles, especially hydroxyapatite has been utilized as carriers for controlled drug delivery due to its non-toxicity and biocompatibility. These particles can encapsulate drugs or therapeutic agents and deliver them in a precise and controlled manner, maximizing the therapeutic benefits while minimizing potential side effects. This controlled release enables the delivery of the encapsulated substances to specific target sites in the body, ensuring localized and sustained drug release.

Due to slow degradation of HA in acidic environment, it is used as a carrier of drugs for treating bone defects like osteoporosis (a major implication after leg surgeries) and bone tumor resection. The inclusion of drugs into porous HA enables precise delivery to the desired area. Antibiotics, anti-cancer drugs and even vitamins, growth hormones, and proteins can be delivered to hard tissues effectively using HA. The local and controlled release of these drugs shortens the time of delivery, accelerates the bone healing process and minimizes the need for extensive surgical intervention in affected bone areas.

2.5.4 GENE DELIVERY

Calcium phosphate nanoparticles have been employed as vehicles to transport genetic material in gene therapy applications. These nanoparticles can encapsulate plasmid DNA or siRNA and efficiently deliver them into target cells, allowing for gene expression modulation or gene silencing.

2.6 DISCUSSIONS

Figure 2.12 shows FTIR and Raman spectra of different forms of calcium phosphate available. Raman studies have confirmed that hexagonal HA has *P63/m* symmetry and a monoclinic unit cell has *P21/b* symmetry. However, the arrangement of OH$^-$

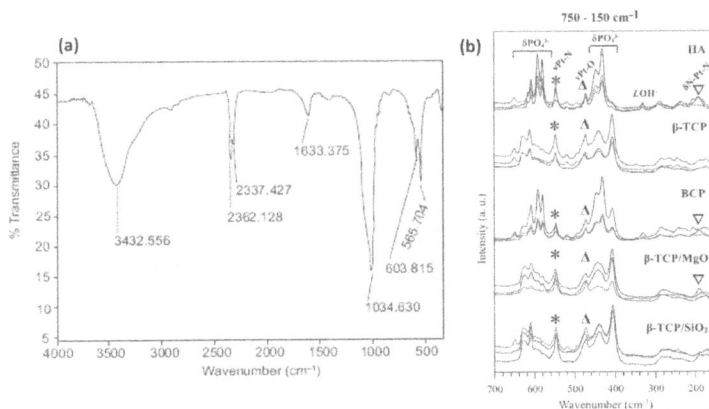

Figure 2.12   (a) FTIR spectra and (b) Raman spectra of calcium phosphates [41, 42]. (Adapted with permission from (a) Granados-Correa Francisco, Juan Bonifacio-Martinez, and Juan Serrano-Gomez. "Synthesis and characterization of calcium phosphate and its relation to Cr (VI) adsorption properties." Revista internacional de contaminacin ambiental 26.2 (2010): 129–134. Copyright 2010. (b) Savicki Cristiane, Nelson Heriberto Almeida Camargo, and Enori Gemelli. "Crystallization of carboplatin-loaded onto microporous calcium phosphate using high-vacuum method: Characterization and release study." PloS ONE 15.12 (2020): e0242565. Copyright 2020.)

anions within the structure associated with the monoclinic and non-orthogonal super-structure, appears to have minimal impact on the vibrational behaviour of the PO_4^{3-} ions within the HA crystal.

The spectra of tricalcium phosphate (TCP) and tetracalcium phosphate (TTCP) exhibit some complexity due to superimposing spectral lines. In α-TCP, the presence of large number of unresolved spectral lines can be attributed to the larger phosphate groups within its structure. Strong lines observed in FTIR spectra of TTCP show similar behavior to oxyapatite indicating the presence of oxide ions in TTCP structures. Some other prominent lines in FTIR spectra are also seen corresponding to vibrations of OH^- in hydroxyapatite and HPO_4^{2-} in octacalcium phosphates (OCP). Raman spectra also shows vibrations at very low wavenumbers.

Overtone and combination bands are seen corresponding to some phases. The overtone/combination band of HA is prominent at 1950–2200 cm^{-1}. The spectral bands for amorphous calcium phosphate (ACP) are broad and asymmetric indicating that it is a mixture of several components.

2.7 CONCLUSION

Calcium phosphate is being widely explored for its applications in biomedical industry. The synthesis of calcium phosphate materials has advanced significantly, offering a wide range of fabrication techniques and strategies to tailor their properties for specific applications.

Various calcium phosphate phases, such as hydroxyapatite, tricalcium phosphate, and amorphous calcium phosphate, have been successfully synthesized and characterized. Each phase possesses unique chemical and physical properties that make them suitable for different biomedical applications. Hydroxyapatite, in particular, has garnered considerable attention due to its remarkable biocompatibility, ability to support bone growth, and similarity to bone matrix. What makes it even more versatile is the inclusion of additives, like nanoparticles, growth hormones, and antimicrobial agents, has expanded the functionalities of calcium phosphate materials, enhancing their therapeutic efficacy and antibacterial properties.

With continued research and technological advancements, calcium phosphate-based materials are poised to revolutionize medical treatments, improve patient outcomes, and pave the way for innovative approaches in regenerative medicine, tissue engineering, and implantable devices.

REFERENCES

1. S. Basu and B. Basu. Unravelling doped biphasic calcium phosphate: Synthesis to application. *ACS Applied Bio Materials*, **2**(12):5263–5297, 2019.
2. A. Szcześ, L. Hołysz, and E. Chibowski. Synthesis of hydroxyapatite for biomedical applications. *Advances in Colloid and Interface Science*, **249**:321–330, 2017.
3. S. Basu and B. Basu. Doped biphasic calcium phosphate: Synthesis and structure. *Journal of Asian Ceramic Societies*, **7**(3):265–283, 2019.

4. R. G. Carrodeguas and S. D. Aza. α-tricalcium phosphate: Synthesis, properties and biomedical applications. *Acta Biomaterialia*, **7**(10):3536–546, 2011.

5. C. Moseke and U. Gbureck. Tetracalcium phosphate: Synthesis, properties and biomedical applications. *Acta Biomaterialia*, **6**(10):3815–3823, 2010.

6. S. Rujitanapanich, P. Kumpapan, and P. Wanjanoi. Synthesis of hydroxyapatite from oyster shell via precipitation. *Energy Procedia*, **56**:112–117, 2014.

7. J. Enax, H.-O. Fabritius, K. Fabritius-Vilpoux, B. T. Amaechi, and F. Meyer. Modes of action and clinical efficacy of particulate hydroxyapatite in preventive oral health care-state of the art. *The Open Dentistry Journal*, **13**:274–287, 2019.

8. X. Lu, H. Zhang, Y. Guo, Y. Wang, X. Ge, Yang Leng, and F. Watari. Hexagonal hydroxyapatite formation on TiO_2 nanotubes under urea modulation. *CrystEngComm*, **13**(11):3741–3749, 2011.

9. J. C. Elliott. *Structure and Chemistry of the Apatites and Other Calcium Orthophosphates*. Elsevier, Amsterdam, 2013.

10. M. Yashima, A. Sakai, T. Kamiyama, and A. Hoshikawa. Crystal structure analysis of β-tricalcium phosphate $Ca_3 (PO_4)_2$ by neutron powder diffraction. *Journal of Solid State Chemistry*, **175**(2):272–277, 2003.

11. M. Yashima and A. Sakai. High-temperature neutron powder diffraction study of the structural phase transition between α and α' phases in tricalcium phosphate $Ca_3 (PO_4)_2$. *Chemical Physics Letters*, **372**(5–6):779–783, 2003.

12. M. Bohner, B. L. G. Santoni, and Nicola Döbelin. β-tricalcium phosphate for bone substitution: Synthesis and properties. *Acta Biomaterialia*, **113**:23–41, 2020.

13. N. Kobayashi, Y. Hashimoto, A. Otaka, T. Yamaoka, and S. Morita. Porous alpha-tricalcium phosphate with immobilized basic fibroblast growth factor enhances bone regeneration in a canine mandibular bone defect model. *Materials*, **9**(10):853, 2016.

14. W. E. Brown and E. F. Epstein. Crystallography of tetracalcium phosphate. *Journal of Research of the National Bureau of Standards. Section A, Physics and Chemistry*, **69**(6):547, 1965

15. A. Y.-Yilmaz and S. Yilmaz. Wet chemical precipitation synthesis of hydroxyapatite (ha) powders. *Ceramics International*, **44**(8):9703–9710, 2018.

16. K. Agrawal, G. Singh, D. Puri, and S. Prakash. Synthesis and characterization of hydroxyapatite powder by sol-gel method for biomedical application. *Journal of Minerals & Materials Characterization & Engineering*, **10**(8):727–734, 2011.

17. J. Chen, Y. Wang, X. Chen, L. Ren, C. Lai, W. He, and Q. Zhang. A simple sol-gel technique for synthesis of nanostructured hydroxyapatite, tricalcium phosphate and biphasic powders. *Materials Letters*, **65**(12):1923–1926, 2011.

18. K. Zhang and K. S. Vecehio. Hydrothermal synthesis of hydroxyapatite rods. *Journal of Crystal Growth*, **308**(1):133, 2007.

19. G. Zhang, J. Chen, S. Yang, Q. Yu, Z. Wang, and Q. Zhang. Preparation of amino-acid-regulated hydroxyapatite particles by hydrothermal method. *Materials Letters*, **65**(3):572–574, 2011.

20. H.-B. Zhang, K.-C. Zhou, Z.-Y. Li, S.-P. Huang, and Y.-Z. Zhao. Morphologies of hydroxyapatite nanoparticles adjusted by organic additives in hydrothermal synthesis. *Journal of Central South University of Technology*, **16**(6):871–875, 2009.

21. B. Yeong, X. Junmin, and J. Wang. Mechanochemical synthesis of hydroxyapatite from calcium oxide and brushite. *Journal of the American Ceramic Society*, **84**(2):465–467, 2001.

22. L. Sinusaite, I. Grigoraviciute-Puroniene, A. Popov, K. Ishikawa, A. Kareiva, and A. Zarkov. Controllable synthesis of tricalcium phosphate (tcp) polymorphs by wet precipitation: Effect of washing procedure. *Ceramics International*, **45**(9):12423–12428, 2019.

23. D. Moreno, F. Vargas, J. Ruiz, and M. E. López. Solid-state synthesis of alpha tricalcium phosphate for cements used in biomedical applications. *boletín de la sociedad española de cerámica y vidrio*, **59**(5):193–200, 2020.

24. G. K. Lim, J. Wang, S. C. Ng, and L. M. Gan. Formation of nanocrystalline hydroxyapatite in nonionic surfactant emulsions. *Langmuir*, **15**(22):7472–7477, 1999.

25. M. Aizawa, T. Hanazawa, K. Itatani, F. S. Howell, and A. Kishioka. Characterization of hydroxyapatite powders prepared by ultrasonic spray-pyrolysis technique. *Journal of Materials Science*, **34**:2865–2873, 1999.

26. G.-H. An, H.-J. Wang, B.-H. Kim, Y.-G. Jeong, and Y.-H. Choa. Fabrication and characterization of a hydroxyapatite nanopowder by ultrasonic spray pyrolysis with salt-assisted decomposition. *Materials Science and Engineering: A*, **449**:821–824, 2007.

27. S. Sasikumar and R. Vijayaraghavan. Synthesis and characterization of bioceramic calcium phosphates by rapid combustion synthesis. *Journal of Materials Science & Technology*, **26**(12):1114–1118, 2010.

28. A. Cüneyt Tas. Combustion synthesis of calcium phosphate bioceramic powders. *Journal of the European Ceramic Society*, **20**(14–15):2389–2394, 2000.

29. S. K. Ghosh, S. Pal, S. K. Roy, S. K. Pal, and D. Basu. Modelling of flame temperature of solution combustion synthesis of nanocrystalline calcium hydroxyapatite material and its parametric optimization. *Bulletin of Materials Science*, **33**:339–350, 2010.

30. L. C. Chow and S. Takagi. Self-setting calcium phosphate cements and methods for preparing and using them, 1996. US Patent 5525148.

31. J. K. Abifarin, D. O. Obada, E. T. Dauda, and D. Dodoo-Arhin. Experimental data on the characterization of hydroxyapatite synthesized from biowastes. *Data in brief*, **26**:104485, 2019.

32. G. A. H. Mekhemer, H. Bongard, A. A. B. Shahin, and M. I. Zaki. FTIR and electron microscopy observed consequences of HCl and CO_2 interfacial interactions with synthetic and biological apatites: Influence of hydroxyapatite maturity. *Materials Chemistry and Physics*, **221**:332–341, 2019.

33. H. Demirkiran, Y. Hu, L. Zuin, N. Appathurai, and P. B. Aswath. Xanes analysis of calcium and sodium phosphates and silicates and hydroxyapatite–bioglass® 45s5 co-sintered bioceramics. *Materials Science and Engineering: C*, **31**(2):134–143, 2011.

34. K.-D. Kwon, J.-S. Chang, S.-H. Lee, D.-H. Lee, K.-S. Lee, J.-H. Hwang, H.-S. Lee, and S.-E. Byun. The effect of cefazolin on mechanical properties and antibacterial reactions of calcium phosphate cement. *Journal of the Korean Orthopaedic Association*, **46**(4):273–281, 2011.

35. T. Qin, X. Li, H. Long, S. Bin, and Y. Xu. Bioactive tetracalcium phosphate scaffolds fabricated by selective laser sintering for bone regeneration applications. *Materials*, **13**(10):2268, 2020.

36. H.-L. Kim, G.-Y. Jung, J.-H. Yoon, J.-S. Han, Y.-J. Park, D.-G. Kim, M. Zhang, and D.-J. Kim. Preparation and characterization of nano-sized hydroxyapatite/alginate/chitosan composite scaffolds for bone tissue engineering. *Materials Science and Engineering: C*, **54**:20–25, 2015.

37. F. J. Martínez-Vázquez, M. V. Cabañas, J. L. Paris, D. Lozano, and M. Vallet-Regí. Fabrication of novel Si-doped hydroxyapatite/gelatine scaffolds by rapid prototyping for drug delivery and bone regeneration. *Acta Biomaterialia*, **15**:200–209, 2015.

38. A. E. Jakus and R. N. Shah. Multi and mixed 3D-printing of graphene-hydroxyapatite hybrid materials for complex tissue engineering. *Journal of Biomedical Materials Research Part A*, **105**(1):274–283, 2017.

39. Q. Yunan, C. Qin, J. Wu, A. Xu, Z. Zhang, J. Liao, S. Lin, X. Ren, and P. Zhang. Synthesis and characterization of cerium-doped hydroxyapatite/polylactic acid composite coatings on metal substrate. *Materials Chemistry and Physics*, **182**:324–332, 2016.

40. P. Cheang and K. A. Khor. Addressing processing problems associated with plasma spraying of hydroxyapatite coatings. *Biomaterials*, **17**(5):537–544, 1996.

41. F. Granados-Correa, J. Bonifacio-Martinez, and J. Serrano-Gomez. Synthesis and characterization of calcium phosphate and its relation to Cr(VI) adsorption properties. *Revista internacional de contaminación ambiental*, **26**(2):129–134, 2010.

42. C. Savicki, N. H. A. Camargo, and E. Gemelli. Crystallization of carboplatin-loaded onto microporous calcium phosphate using high-vacuum method: Characterization and release study. *Plos One*, **15**(12):e0242565, 2020.

3 Calcium-Based Magnetic Biochars

Sanjeev Gautam and Ruhani Baweja
Advanced Functional Materials Laboratory, Dr. S.S. Bhatnagar University Institute of Chemical Engineering and Technology, Panjab University, Chandigarh, India

3.1 INTRODUCTION

With the rapid increase in population there is increase in demand for the basic necessities, this lead to rise in different types of industries. But this increase has various negative impacts on the environment. To elaborate, the agricultural waste is increasing day by day and it became very difficult to dispose of the waste. Some countries like Malaysia generate a significant amount of agricultural waste, with an annual production exceeding two million tons. This agricultural waste includes various by-products and residues from the country's agricultural sector, such as rice straw, palm oil biomass, fruit peels, and other plant residues [1]. These wastes are dumped in landfill or released into water, which creates soil, water, and air pollution and reduce the available resources for the society. These agricultural wastes contain minerals such as silica, magnesium, and potassium which are highly porous. Apart from this, there are certain materials like arsenic and cadmium which are very harmful for the human beings. These materials are used in electroplating and battery manufacturing and are produced by mining activities and fossil fuel combustion and when these harmful elements are released into the environment, they cause water and soil pollution [2]. Cadmium have severe effects on the human body such as it can damage the brain, bones and even liver as it is highly toxic material because of its non-biodegradable nature. To remove Cd from soil various techniques have been launched till date such as chemical precipitation, soil washing etc. However, these technologies may be disruptive to the environment [3]. Malachite green (MG) is a widely used water-soluble cationic dye that belongs to the class of triphenylmethane-based synthetic refractory organic compounds and having many stable chemical properties, which promote its use in paper and textile industries, also antibacterial and antifungal agent. But it is a highly carcinogenic, i.e., it can causes cancer and genotoxic, which means it can damage genetic material. It can also lead to harm and impairment of the immune system's normal functioning, which is responsible for protecting the body against infections and diseases [4].

DOI: 10.1201/9781003360599-3

Table 3.1

Properties of Biochar and Their Function and Their Applications in Pollutant Absorption [13]

Biochar Property	Function	Application/ Mechanism Involved
Specific surface area (SSA), porosity	Sorption capacity	Immobilization of contamination from solid, liquid and gaseous media
Ion-exchange properties on surface	Electrostatic interactions with polar or non-polar groups	Adsorption of organic contaminants
Surface functional groups	Chemical bonding with certain molecules	Absorption and immobilization of certain toxins and drugs

So, it becomes necessary to control the various effluents which are harmful for environment and human beings. Hence, there is need to devise a method which is cost effective and does not affect the environment further. Biochar is a type of solid material that is rich in carbon and is produced through a process known as pyrolysis. Pyrolysis involves heating biomass, such as wood, agricultural residues, or organic waste, in the absence of oxygen or with limited oxygen supply. This condition, called hypoxia or anaerobic, prevents complete combustion of the biomass and instead leads to the thermal decomposition of organic matter [5]. Biochar possesses several distinct characteristics that make it a valuable material with a wide range of applications. First, it contains a large number of functional groups. These functional groups are chemical moieties or reactive sites on the surface of the biochar that can interact with other substances. They can include carboxyl, hydroxyl, phenolic, and aromatic groups. Second, biochar exhibits a significant surface area [6]. The pyrolysis process creates a porous structure within the biochar, resulting in a high surface area-to-volume ratio. This increased surface area provides more sites for chemical reactions, adsorption, and microbial colonization. The porous structure allows for the retention of water, nutrients, and beneficial microorganisms, promoting better soil fertility and moisture retention. These properties help in absorbing heavy metals from the soil and control soil pollution. It reduces the heavy metal contamination by reducing the mobility of heavy metals. Some of properties of biochar along with their application are discussed in Table 3.1. It was found that certain heavy metal pollutants are not appropriately adsorbed by the biochar and hence lot of modifications are developed such as ultraviolet modification, acid/base treatment which increases the remediation efficiency of biochar [7]. Due to its negatively-charged surface, it has strong absorption ability for cationic heavy metals like Cd(II) but less for anionic ions like As.

In contrast, Fe oxide-based materials have higher attraction for As because it form chelate compounds with oxides. Hence, combining Fe oxide with biochar will help in forming a biochar material which helps in the contamination of heavy metals [8]. Magnetic biochars can be prepared by various methods, for instance, in laboratory it can be prepared by conventional heating in an electric furnace [9], co-precipitation method [10], calcination method [11], etc. It was found that the biochar which is manufactured from the plant waste and animal residue can be used to adsorb organic pollutants, in contrast, the biochar which is derived from the animal manure can be used to adsorb both organic and metallic contaminants [12].

The remediation efficiency of fields containing arsenic (As) can be enhanced through the combination of iron oxide and calcium carbonate with biochar, known as Ca-MBC. Iron oxide possesses an improved attraction capability for arsenic, while calcium carbonate ($CaCO_3$) helps neutralize the residual acidity of the materials. This is achieved by incorporating ferrous sulphate and ferric chloride. So, the addition of calcium compounds helps in the maintenance of soil pH. Additionally, the magnetic properties of the material enable easy removal through the application of an external magnetic field, making it suitable for re-utilization and preventing secondary pollution [14].

Magnetic biochar have solved the various problems by combining biochar with some magnetic ions. These help in absorbing waste water effectively and helps in the removal of harmful pollutants from water. Magnetic biochars are also to make the electrodes of supercapacitors and helps in increasing the electrical conductivity and capacitance [15]. Another reason for the introduction of magnetic ions is that during the process some of the pollutants may get desorbed from the surface of biochars, which as a result creates secondary pollution. Hence, some transition metals like Fe, Co, etc. or some oxides are added within the biochars matrices to avoid desorption and results in formation of magnetic biochars. These are also useful in cleaning nuclear waste polluted water [16]. The applications of magnetic biochars specifically depend upon the additional raw material added to the biochar and upon the synthesis methods used for its preparation and parameters such as temperature, etc. [17].

3.2   MODIFICATION OF BIOCHAR WITH CALCIUM

Magnetic biochars which are obtained by the combination of biomass with ferrites using a pyrolysis technique can act as a good adsorbent, but its adsorbing properties can be further improved by the surface modification techniques. Many materials like Cu, Mg, etc. [18, 19] having high pollutant adsorbing tendency can be used. Calcium is quite abundant material having non-toxic nature, so, it is highly preferred as a surface modifier [20].

The addition of calcium in biochar helps in the reduction of soil cation exchange capability as shown in Figure 3.1, i.e., the negative charges in the soil and hence increases the strongest adsorption abilities for cadmium. It can absorb and make complexes of the As in the soil which increases the residual fraction of arsenic in soil [21]. The addition of calcium can modify the pH level and made the biochar more alkaline. The formation of insoluble metal-calcium complexes on the biochar surface can increase the retention of heavy metals. Calcium ions can compete with metal ions for binding sites on the biochar, resulting in the immobilization of heavy metals [8].

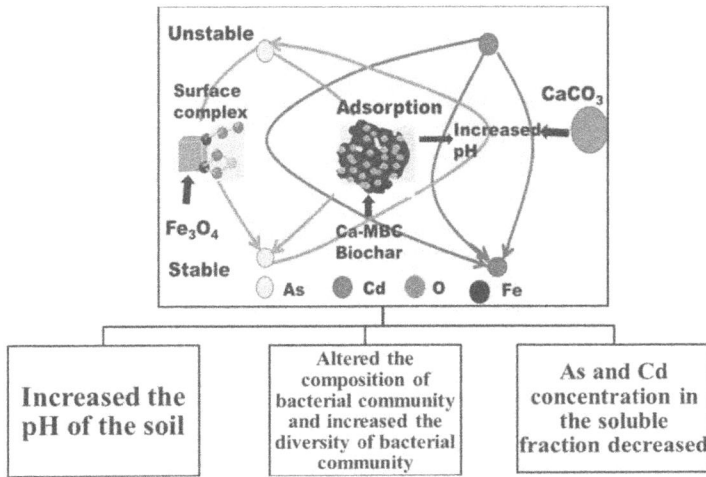

Figure 3.1 Applications of calcium-based magnetic biochar in co-contamination of heavy metals from soil [8]. (Adapted from Wu *et al. J. Hazard. Mater.*, 387:122010, 2020. Copyright 2020 Elsevier.)

3.3 SYNTHESIS METHODS

The limitation of biochar is that it gets attached to aqueous solution strongly and hence difficult to separate and require techniques like centrifugation, sedimentation and filtration and are costly. So to overcome this bottleneck problem magnetic materials are incorporated inside the biochar like Fe, since, they have stronger tendency for the separation techniques. The addition of magnetic materials affect the physicochemical properties of biochar and both these materials act together to remove the pollutants from aqueous solution [17]. This helps in the aggregation of photocatalysts which can appropriately disperse on the surface of magnetic biochars. These also act as adsorbents for the organic pollutants [22]. It helps in removal of heavy metal contaminants, polycyclic aromatic hydrocarbons, steroids, etc. To add more, magnetic biochars facilitate plant growth, improve fertility of soil and act as useful component for the environment. In addition to its slow-release fertilizer properties, biochar also plays a role in regulating the alkalinity of the soil. The pH value of biochar, which is typically neutral to slightly alkaline, can help balance and stabilize soil pH. In acidic soils, biochar acts as a pH buffer, raising the pH toward a more neutral level. [13]. The magnetic biochars has multiple applications due to which it is in high demand these days, as it can be used in carbon storage capacity, toxins and pollutants absorber, for the reduction of greenhouse gases, in agricultural applications, etc. [12]. Different methods are used for the preparation of magnetic materials like co-precipitation method, pyrolysis, hydrothermal method and calcination, these are describes in Table 3.2 with their advantages and disadvantages.

Table 3.2

Description of Preparation Methods of Magnetic Biochar, Their Advantages and Their Disadvantages [22]

Preparation Method	Advantages	Disadvantages
Pyrolysis	<ul><li>Operation is simple</li><li>MBC can be prepared by one step method</li><li>Good stability and less metal leaching</li></ul>	<ul><li>Gas pollutants and tar will be produced</li><li>Need higher temperature</li></ul>
Co-precipitation	<ul><li>Excellent stability and less metal leaching</li><li>No excessive temperature required</li><li>High productivity and strong controllability</li></ul>	<ul><li>Large number of alkaline reagents needed</li><li>High cost</li><li>Alkaline waste water need to be treated.</li></ul>
Hydrothermal method	<ul><li>Reaction is mild</li><li>Does not need high temperature</li><li>No need for dehydration</li></ul>	<ul><li>Stability of MBC is worse</li><li>Need to introduce alkaline reagents, reductants or surfactants to improve the stability of MBC</li></ul>
Reductive co-precipitation	<ul><li>Can get MBC loaded with zero-valent metal</li><li>Strong reducibility and good stability</li><li>High production efficiency</li><li>Strong controllability</li></ul>	<ul><li>Added reductants are toxic</li><li>Hydrogen is usually generated during the preparation process, which has some potential safety risks</li></ul>

3.3.1   CO-PRECIPITATION METHOD

In this method, the biochar is mixed in a solution having transition metals and then sodium hydroxide and ammonium hydroxide is added to it at a certain temperature with continuous stirring at a pH 9–11, followed by the removal of supernatant, after which residue is left which can be dried to form magnetic biochar [23]. The schematic diagram representing the procedure of co-precipitation method is shown in Figure 3.2.

$$Fe^{2+} + 2Fe^{3+} + 8OH^- \rightarrow Fe_3O_4 + 4H_2O \tag{3.1}$$

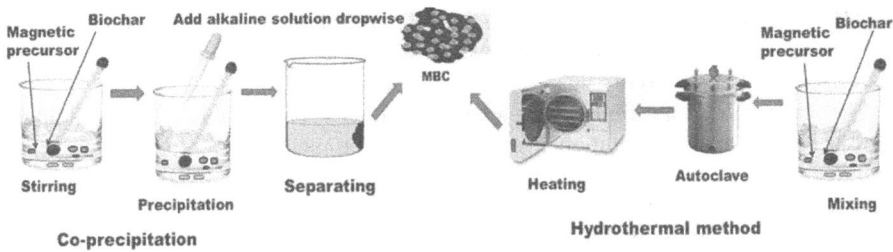

Figure 3.2   Co-precipitation method and Hydrothermal method of synthesis of magnetic biochar.

3.3.2   HYDROTHERMAL CARBONIZATION

In this method of synthesis, heterogeneous reaction takes place between transition metal ions and biomass, at low reaction temperature which is accompanied by pouring the solution in an autoclave. The pressure is generated by the reaction itself. As no base or strong reductants are added, so it is easier than other methods [24]. The schematic diagram representing the procedure of hydrothermal method is shown in Figure 3.2.

3.3.3   REDUCED CO-DEPOSITION METHOD

The method described is similar to the co-precipitation method, but it incorporates an additional step involving the reduction of the transition metal with a reducing agent like sodium borohydride, during the binding process to biochar. Following this step, the supernatant is removed, and the remaining residue is dried to obtain magnetic biochar. The advantage of this method lies in the resulting magnetic biochar's high reductive capacity and suitability for contaminant removal. The combination of the transition metal and biochar creates a composite material with enhanced contaminant adsorption and reduction capabilities [25, 26].

$$2Fe^{3+} + 6BH_4^- + 18H_2O \rightarrow 2Fe^\circ + 6B(OH)_3 + 21H_2 \qquad (3.2)$$

3.4   SYNTHESIS OF CALCIUM BASED MAGNETIC BIOCHAR

The biomass which is used for the preparation of biochar mainly include two types, one of which comes from crop wastes such as wheat straw, corn straw, rice husks, potato stems, cotton straw, green tea residue [27], and likewise. This is most prominently used, as it is readily available. Another one containing carbon wastes including algae, shells, anaerobic digestion residue, common sludge, animal fecal and animal bone etc. [28]. Since the above biomass contains very less amount of iron content, so in this magnetic precursors are added which results in the formation of magnetic biochar. The magnetic precursors which are added to this mainly consists of three types that are transition metal salts like $FeSO_4 \cdot nH_2O$, $FeCl_3 \cdot nH_2O$,

$Fe(NO_3)_3 \cdot nH_2O$, etc., natural iron ores such as hematite, pyrite, magnetite etc. and iron oxides such as Fe_2O_3 and Fe_3O_4. Among these transition metals and iron oxides are expensive but are highly pure [29].

Pyrolysis acts as an imperative technology to treat animal manures and help in killing pathogens and decomposing antibiotics. The method offers several benefits in the conversion of organic matter found in animal manures into value-added products such as biochar, biogas, and bio-oil. This process allows for the transformation of waste materials into useful resources, promoting sustainability and resource efficiency. Furthermore, the method facilitates the immobilization of heavy metals within the produced biochar [30]. The typical diagram representing the synthesis of magnetic biochar is shown in Figure 3.3. Heavy metals, which can be present in animal manures due to various factors such as animal feed or environmental contamination, pose a risk to ecosystems and human health. However, during the pyrolysis or anaerobic digestion process, the heavy metals present in the manures become trapped and bound within the biochar structure. This immobilization helps in their removal from the immediate environment, reducing their potential adverse effects. Different types of calcium products like CaO, $CaCO_3$, $Ca(OH)_2$, and $Ca(H_2PO_4)_2$ are incorporated in biochars and their characteristics are determined [31]. The results showed that the more suitable material for improving the properties of the biochar is CaO. It was found that on the addition of calcium carbonate, more number of gaseous and liquid products are converted, on the other hand, by the addition of CaO, H_2 and CH_4 production is favoured while CO_2 emission is reduced and this affects the various properties of biochar which includes its elemental composition, surface properties, morphology, and functional groups. $Ca(H_2PO_4)_2$ affect the carbon monoxide and carbon dioxide yield but have no effect on hydrogen and methane [32].

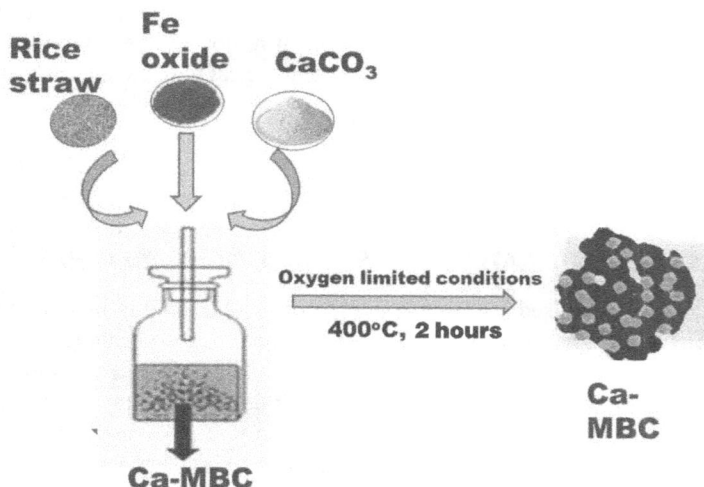

Figure 3.3 Preparation of calcium based magnetic biochar.

Thus, one of the methods of incorporating calcium inside the biochar is the pyrolyzation method. In this method, the mixture of rice straw, Fe oxide, and calcium carbonate is pyrolyzed. The dried rice straw is added into the solution which contains ferrous and ferric ions. Additionally, the solution is adjusted using sodium hydroxide to pH 12 and the suspension is stirred continuously. After desiccating it, calcium carbonate was added to it and the mixture was then pyrolyzed to obtain calcium based magnetic biochar,in controlled atmosphere furnace [2].

The process can be best described by the following reactions $\rightarrow$

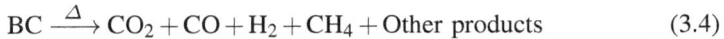

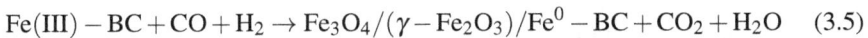

$$Fe^{3+} + BC \rightarrow Fe(III) - BC \tag{3.3}$$

$$BC \xrightarrow{\Delta} CO_2 + CO + H_2 + CH_4 + \text{Other products} \tag{3.4}$$

$$Fe(III) - BC + CO + H_2 \rightarrow Fe_3O_4/(\gamma - Fe_2O_3)/Fe^0 - BC + CO_2 + H_2O \tag{3.5}$$

3.5 CHARACTERIZATION

Many techniques can be used to characterize the biochar formed which are described in Figure 3.4 and some of them are discussed here.

3.5.1 DETERMINATION OF MORPHOLOGY OF MAGNETIC BIOCHAR

The morphology of calcium based magnetic biochar can be determined by using SEM and TEM analysis techniques as shown in Figure 3.5. It is seen clearly visible from SEM results that diamond shaped blocks are attached on the surface of MBC, while rectangular-shaped substances are seen in Ca/MBC. From TEM results, it is observed that in MBC spherical substances are attached and in Ca/MBC large size regular substances are seen [4]. It is also observed that the presence of the transition metal salt will lead to the increase in the blocking of the pores of metal biochars as a

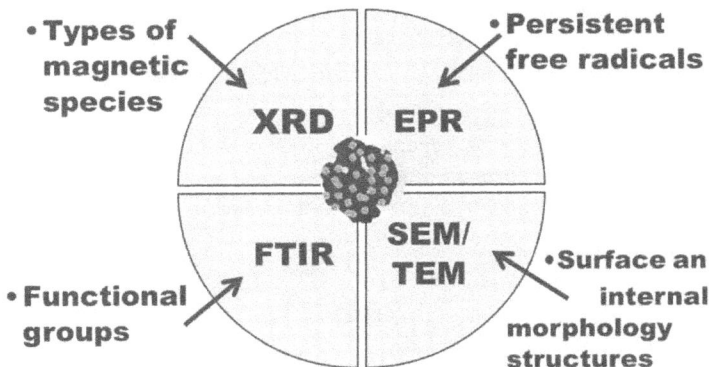

Figure 3.4 Various techniques to characterize Magnetic biochar.

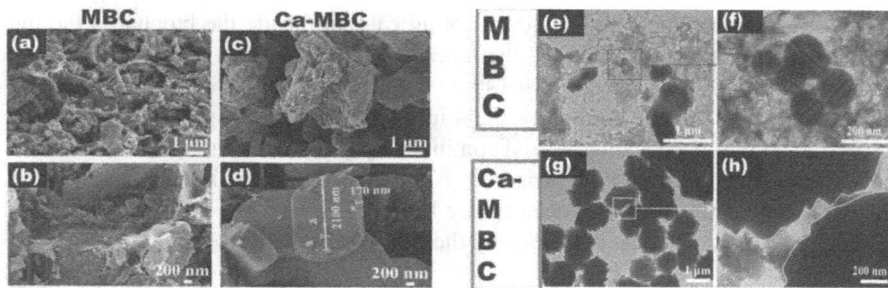

Figure 3.5 (a,b) SEM of Magnetic Biochar. (c,d) SEM of Calcium based magnetic Biochar. (e,f) TEM of Magnetic Biochar. (g,h) TEM of Calcium based magnetic Biochar [4]. (Adapted with permission under an open access by CC from Wang *et al. Int. J. Env. Pub. Health*, 19:6, 2022. Copyright 2022 MDPI.)

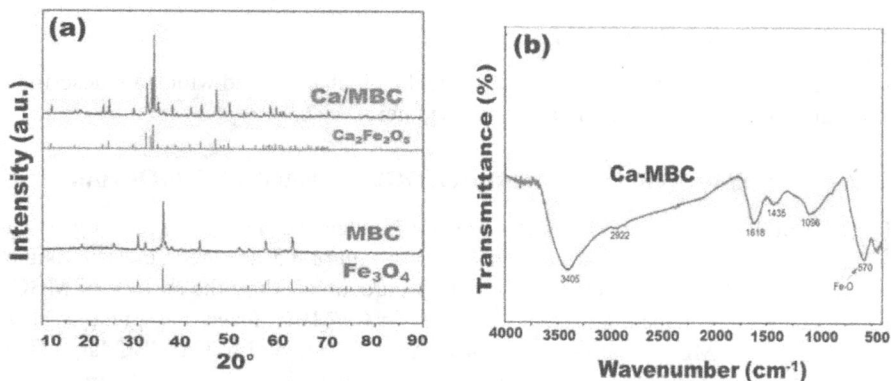

Figure 3.6 (a) XRD pattern of magnetic biochar having Fe_3O_4 and calcium based magnetic biochar having $Ca_2Fe_2O_5$ [4] . (Adapted with permission under an open access by CC from wang *et al. Int. J. Env. Pub. Health*, 19:6, 2022. Copyright 2022 MDPI.) (b) FTIR analysis of calcium based magnetic biochar to detect the various functional groups present [2]. (Adapted from Wu *et al. J. Hazard. Mater.*, 348:10–19, 2018. Copyright 2018 Elsevier.)

result of this, specific surface area decreases [33]. The XRD pattern of MBC is found to contain different peaks corresponding to Fe_3O_4 as shown in Figure 3.6 (a).

The reaction between biochar and ferrate ions occur by following pathway →

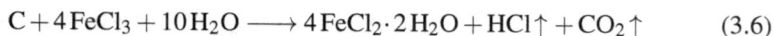

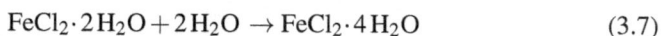

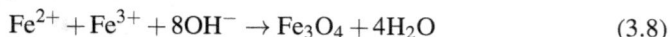

$$C + 4\,FeCl_3 + 10\,H_2O \longrightarrow 4\,FeCl_2 \cdot 2\,H_2O + HCl\uparrow + CO_2\uparrow \tag{3.6}$$

$$FeCl_2 \cdot 2\,H_2O + 2\,H_2O \rightarrow FeCl_2 \cdot 4\,H_2O \tag{3.7}$$

$$Fe^{2+} + Fe^{3+} + 8\,OH^- \rightarrow Fe_3O_4 + 4\,H_2O \tag{3.8}$$

On the addition of calcium large number of peaks appears in the spectrum corresponding to $Ca_2Fe_2O_5$, in addition to MBC peaks. It can be inferred that the spherical substance attached to the surface of magnetic biochar may be Fe_3O_4 and the rectangular structured type of substance on Ca/MBC is $Ca_2Fe_2O_5$ [4].

3.5.2 SURFACE AREA AND POROUS STRUCTURE

The surface area of MBC increases with increase in the temperature as because of formation of more pores which may be because of the more release in the volatile matter at high temperature. In addition, with the increase in calcium containing compounds the surface area increases further [20].

3.5.3 FUNCTIONAL GROUPS

The FT-IR spectroscopy analysis of magnetic biochar reveals specific peaks at different wavenumbers, indicating the presence of certain functional groups and molecular vibrations.

- At 570 and 441 cm^{-1}, peaks are observed, which correspond to the FeO stretching vibration of high-spin Fe^{3+} complex within the Fe_3O_4 structure. This indicates the presence of iron oxide (Fe_3O_4) in the magnetic biochar. The FeO stretching vibration is a characteristic feature of the iron oxide structure and confirms the presence of the magnetic component in the biochar composite.
- Another peak is observed at 2922 cm^{-1}, which is attributed to the alkyl CH_2 stretching. This peak indicates the presence of hydrocarbon chains, particularly in the form of alkyl groups, within the biochar structure. The alkyl CH_2 stretching vibration is associated with the stretching of carbon-hydrogen bonds in the methylene (-CH_2-) group, which is commonly found in organic compounds.
- Furthermore, peaks are observed at 3405 and 1096 cm^{-1}. The peak at 3405 cm^{-1} is assigned to the stretching vibration of the OH group, indicating the presence of hydroxyl (-OH) functional groups in the biochar. The OH group is typically associated with hydrophilic characteristics and can play a role in the biochar's interactions with water and other polar substances.
- The peak at 1096 cm^{-1} is attributed to the C-O stretching vibration, specifically the stretching of carbon-oxygen bonds in the C-O functional groups. This peak signifies the presence of carbon-oxygen bonds, such as those found in hydroxyl groups (-OH) or carbonyl groups (C=O), within the biochar structure. This is shown in Figure 3.6 (b) [2].

The magnetic biochar of good performance can be obtained by understanding the nature of the functional groups present on the surface of magnetic biochar and their mechanisms of adsorption of the pollutants. It was found that there are free radicals on the surface of magnetic biochar which help in the removal of several pollutants [34].

3.6   ENVIRONMENTAL APPLICATIONS OF CALCIUM BASED MAGNETIC BIOCHARS IN WATER AND SOIL REMEDIATION

Calcium-based magnetic biochar, with its combined calcium modification and magnetic properties, holds great potential for various environmental applications. The unique characteristics of calcium-based magnetic biochar enable its utilization in a range of environmental processes, including pollutant remediation, water treatment, soil fertility enhancement, and sustainable agriculture.

3.6.1   HEAVY METAL REMOVAL

The heavy metals present in the environment are classified as cationic which includes Cd(II), Pb(II) etc. and anionic such as Cr(VI) and arsenic (As(V) and As(III)). The applications of the magnetic biochar depends on the method of synthesis and materials used. Biochar possesses a range of characteristics that make it effective in the removal of heavy metal ions. These characteristics include electrostatic adsorption, reduction, ion exchange, complexation with functional groups, and coprecipitation [17] as shown in Figure 3.7.

- First, electrostatic adsorption occurs when the charged surface of biochar attracts and binds heavy metal ions through electrostatic interactions. The surface charge of biochar, which can be positive or negative depending on the pH conditions, plays a vital role in this process.
- Second, biochar has the capability to facilitate reduction reactions. Some forms of biochar contain carbonaceous materials with strong reducing properties. These materials can interact with heavy metal ions, leading to their reduction to less toxic or less mobile forms.

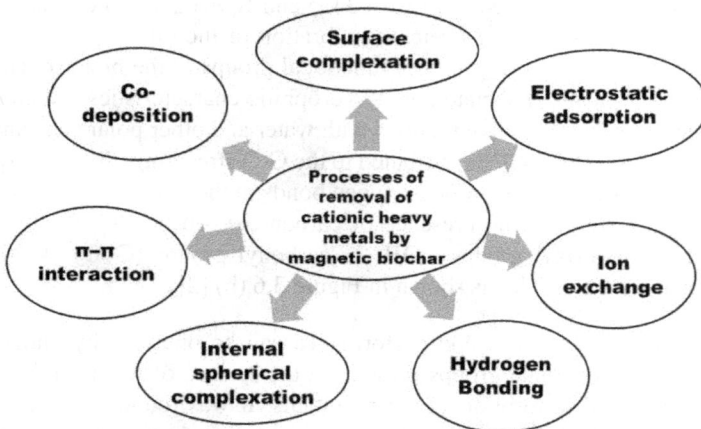

Figure 3.7   Mechanisms by which magnetic biochar interact with cationic heavy metals.

- Biochar exhibits ion exchange properties. This means that heavy metal ions can be exchanged with other ions present on the biochar surface. The exchange occurs when heavy metal ions are attracted to the surface and replace other ions through a chemical process.
- Functional groups present on the surface of biochar can form complexes with heavy metal ions. These functional groups, such as carboxyl, hydroxyl, and phenolic groups, have the ability to bind with metal ions through coordination chemistry, forming stable complexes that prevent their mobility or toxicity.
- Lastly, co-precipitation is another mechanism through which biochar can remove heavy metal ions. When certain conditions are met, heavy metal ions can undergo precipitation reactions along with other compounds present in the biochar, resulting in the formation of insoluble compounds that can be easily separated from the solution.

The heavy metal which is present in soil can also be categorized on different basis such as their reducible state, which is less stable under redox conditions, residue state, oxidized state, not easily adsorbed by the plants and weak extraction state, which can be easily adsorbed by the plants and is highly toxic from them. The addition of biochar helps in the conversion of these states to oxidizable state [7].

To increase the removal of cadmium from the material, the $CaCO_3$ should be added to the biomass-iron oxide mixture but with increase in the content of the calcium carbonate the pH level of the material increases which assists the removal of cadmium but reinforce the removal of arsenic [35]. The addition of calcium based magnetic biochar helps in the increase of residual fraction of arsenic in the soil and this is done by the formation of strong bonding between Fe/Mn oxyhydroxide fraction of As and Ca-MBC, as arsenic can form mono or bi-dentate chelate with the iron [8].

3.6.2 ORGANIC CONTAMINANT ADSORPTION

Magnetic biochar is a type of biochar that possesses magnetic properties, making it useful as a catalyst in the degradation of organic pollutants. It has been observed to exhibit excellent catalytic performance in a wide range of systems. Some of the systems in which biochar can act as good catalyst are peroxydisulfate systems, PMS, Fenton-like systems , photocatalytic, and $NaBH_4$ systems as shown in Figure 3.8.

1. Peroxydisulfate (PS) systems
   In the biochar, when magnetic substance is added then it can effectively affect the peroxydisulfate systems, for instance, Fe_3O_4 attached on MBC then it will release Fe^{2+} ions by the conversion of Fe^{3+} to Fe^{2+}, which are very effective in activating PS systems. Some other metals on MBC can also help in the activation of PS, which can be clear from the following reactions [36]:

$$Fe^{2+} + S_2O_8{}^{2-} \rightarrow Fe^{3+} + SO_4{}^{\bullet-} + SO_4{}^{2-} \qquad (3.9)$$

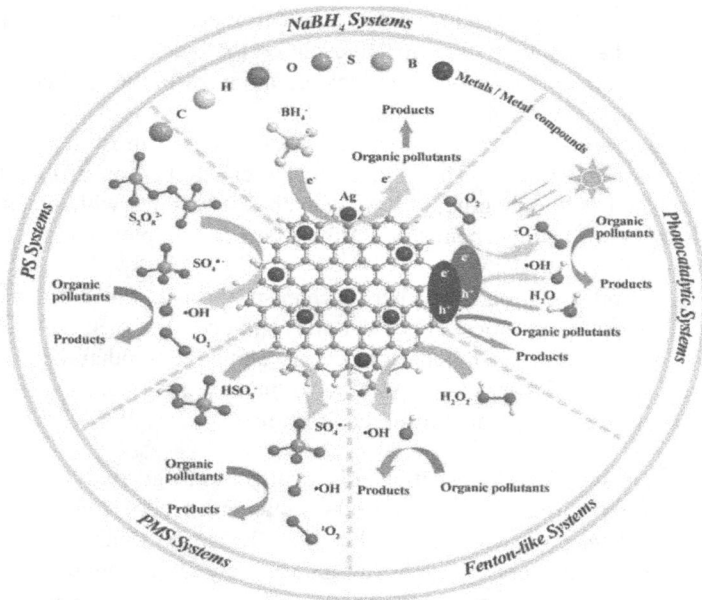

Figure 3.8 Magnetic biochars as a catalyst in various types of systems [22]. (Adapted from Feng *et al.*, *Waste Management*, 123:69–79, 2021. Copyright 2021 Elsevier.)

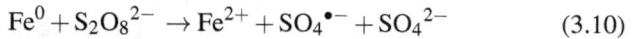

$$Fe^0 + S_2O_8^{2-} \rightarrow Fe^{2+} + SO_4^{\bullet-} + SO_4^{2-} \qquad (3.10)$$

$$Fe^{2+} + S_2O_8^{2-} \rightarrow Fe^{3+} + SO_4^{\bullet-} + SO_4^{2-} \qquad (3.11)$$

$$Ce^{3+} + S_2O_8^{2-} \rightarrow Ce^{4+} + SO_4^{\bullet-} + SO_4^{2-} \qquad (3.12)$$

$$Mn^{2+} + S_2O_8^{2-} \rightarrow Mn^{3+} + SO_4^{\bullet-} + SO_4^{2-} \qquad (3.13)$$

$$Mn^{3+} + S_2O_8^{2-} \rightarrow Mn^{4+} + SO_4^{\bullet-} + SO_4^{2-} \qquad (3.14)$$

MBC contain various functional groups and defects which make it favourable to act as a catalyst in peroxydisulfate systems as shown in Figure 3.9 (a).

- Magnetic biochars contain hydroxyl and carboxyl groups that help in the generation of radicals by transferring electrons to the peroxydisulfate system [37].

$$BC_{surface} - OOH + S_2O_8^{2-} \rightarrow BC_{surface} - OO^\bullet + SO_4^{\bullet-} + HSO_4^- \qquad (3.15)$$

$$BC_{surface} - OH + S_2O_8^{2-} \rightarrow BC_{surface} - O^\bullet + SO_4^{\bullet-} + HSO_4^- \qquad (3.16)$$

- This electron transfer process also helps in the decomposition of PS to sulfate radicals due to the presence of PFRs on the MBC which can act like redox centers [40].

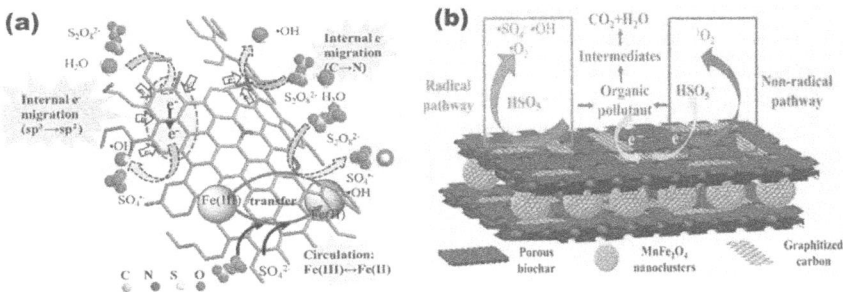

Figure 3.9 (a) N-doped sludge MBC having three active sites of PS which are ironcompounds, graphitic carbon, and doped nitrogen species with mechanism [38]. (Adapted from Yu *et al. J. Chem. Eng.*, 364:149–159, 2019. Copyright 2019 Elsevier.) (b) Mechanism of activation of PMS on $MnFe_2O_4$@BC by two pathways i.e., radical and non-radical pathway [39]. (Adapted from Fu *et al. J. Chem. Eng.*, 360:157–170, 2019. Copyright 2019 Elsevier.)

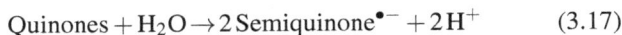

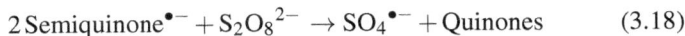

$$Quinones + H_2O \rightarrow 2\,Semiquinone^{\bullet-} + 2\,H^+ \qquad (3.17)$$

$$2\,Semiquinone^{\bullet-} + S_2O_8{}^{2-} \rightarrow SO_4{}^{\bullet-} + Quinones \qquad (3.18)$$

- The defects present in the MBC also helps in the activation of PS. To elaborate, the PS molecules has O-O bond present in it because of linkage SO_4–SO_4 and defects present in MBC will reduce the chemical bond energy of these bonds and make it unstable and consequently, the electron transfer process takes place and leads to the formation of radicals. Thus, magnetic biochar can act as an electron donor and at the same time it helps in the transfer of electrons [40].

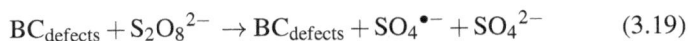

$$BC_{defects} + S_2O_8{}^{2-} \rightarrow BC_{defects} + SO_4{}^{\bullet-} + SO_4{}^{2-} \qquad (3.19)$$

2. Peroxymonosulfate (PMS) systems

To activate PMS systems, MBC uses two methods:
- Radical pathway
- Non-radical pathway

These methods are shown in Figure 3.9 (b).

Radical pathway: In this method, peroxymonosulfate is activated to form radical species like $SO_4^{\bullet-}$ and $\cdot OH$ by the magnetic substances present on magnetic biochars, for instance, iron and cobalt due to its good catalytic properties, helps in the activation of PMS. During this process, iron or cobalt present on MBC then release Fe^{2+} or Co^{2+} slowly, which help in the catalyzation of PMS by series of reaction [41].

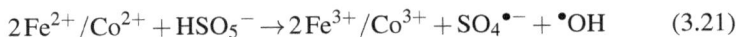

$$Fe^{2+}/Co^{2+} + HSO_5{}^- \rightarrow Fe^{3+}/Co^{3+} + SO_4{}^{\bullet-} + OH^- \qquad (3.20)$$

$$2\,Fe^{2+}/Co^{2+} + HSO_5{}^- \rightarrow 2\,Fe^{3+}/Co^{3+} + SO_4{}^{\bullet-} + {}^\bullet OH \qquad (3.21)$$

Non-radical pathway: The presence of graphite structure in MBC make available large number of positively charged sites on the surface of MBC, which attract PMS and make it adsorbed on surface easily and as a result, produces electron transfer intermediates. These electron transfer intermediates help in the decomposition of organic compounds and graphite structure in this act like a bridge [42]. PMS can also undergo self-decomposition in the presence of magnetic biochar and result in singlet oxygen.

$$HSO_5^- + SO_5^{2-} \rightarrow SO_4^{2-} + HSO_4^- + 1O_2 \tag{3.22}$$

$$HSO_5^- + O_2^{\bullet-} \rightarrow SO_4^{\bullet-} + OH^- + O_2 \tag{3.23}$$

3. Fenton-like systems

For the formation of Fenton-like systems, there is a limit on pH value but with the help of MBC containing iron it is easy to form heterogeneous Fenton-like system with H_2O_2, over a wide range of pH [43]. In this, Fe^{2+} ion react with H_2O_2 and activate it to produce ·OH [44].

$$Fe^{2+} + H_2O_2 \rightarrow Fe^{3+} + OH^- + {}^\bullet OH \tag{3.24}$$

$$Fe^{3+} + H_2O_2 \rightarrow Fe^{2+} + H^+ + HO_2^\bullet \tag{3.25}$$

$$Fe^{3+} + HO_2^\bullet \rightarrow Fe^{2+} + H^+ + O_2 \tag{3.26}$$

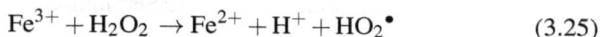

The concentration of MBC and H_2O_2, pH and temperature affect the catalytic properties of the MBC/H_2O_2 system. It was found that the light also affects these systems and improves the catalytic efficiency, since the introduction of UV-vis-light irradiation favours the decomposition of H_2O_2 into ·OH and also promotes the reduction of Fe^{3+} to Fe^{2+} [45].

$$H_2O_2 + h\nu \rightarrow 2\,{}^\bullet OH \tag{3.27}$$

$$Fe^{3+} + H_2O \rightarrow FeOH^{2+} + H^+ \tag{3.28}$$

$$FeOH^{2+} + h\nu \rightarrow Fe^{2+} + {}^\bullet OH \tag{3.29}$$

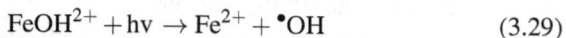

4. Photocatalytic systems

The development of photocatalysts suffer various limitations, for instance, some electron-hole pairs which are generated by light may combine at high rate and thus results in agglomeration and as a result, their production become difficult [46]. Since, the iron oxides have a small band gap, so they form a heterojunction with the photocatalysts, which help in separating electron-hole pairs and thus helps in increasing the efficiency of photocatalysts [47].

The interaction between MBC and photocatalysts occur by the following steps:

• The photoabsorption of heterojunction catalyst.
• Electron-hole pairs are generated by light.

- The transfer of charge carriers.
- The utilization of charge carriers by the reactant [48].

5. NaBH$_4$ systems

   NaBH$_4$ is used for the degradation of organic pollutants and to catalyze it silver nanoparticles are used but these nanoparticles aggregate during the catalysis process. To avoid the aggregation of these nanoparticles, the Ag-NPs are loaded on MBC. Moreover, due to the magnetic property of MBC, it helps in the recovery of catalyst. These systems are used for the degradation of dyes. Here silver nanoparticles are used for the purpose of electron transfer and electrons are donated by BH$_4^-$ and accepted by dye molecules from AgNPs [49].

3.6.3   WATER TREATMENT

There are many factors that influence the efficiency of absorption of pollutants from the water by the magnetic biochar as shown in Figure 3.10 and also discussed further:

1. **Chemical impregnation ratio**

   The ratio of raw material to that of magnetic material influences the adsorption capacity of magnetic biochars. For instance, it was determined in [50] that the adsorption capacity increases with increase in impregnation ratio from 0.25 to 0.50, but for value higher than 0.5 the adsorption capacity decreases, i.e., at 0.5, MBC has the optimum value of adsorption capacity.

2. **Pyrolytic temperature**

   Activation temperature helps in the improvement of the porous structure of the magnetic biochar. For instance, by increasing the pyrolysis temperature, the tendency of carbonation of magnetic biochar increases and the magnetic

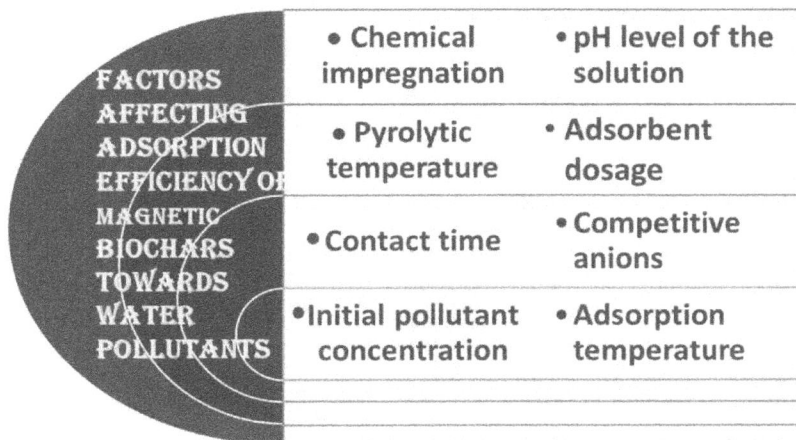

FACTORS AFFECTING ADSORPTION EFFICIENCY OF MAGNETIC BIOCHARS TOWARDS WATER POLLUTANTS	• Chemical impregnation	• pH level of the solution
	• Pyrolytic temperature	• Adsorbent dosage
	• Contact time	• Competitive anions
	• Initial pollutant concentration	• Adsorption temperature

Figure 3.10   Factors on which water pollutants adsorption efficiency depends.

biochars at pyrolysis temperature of 873 K are found to have maximum surface area and the pore volume [51].

3. **pH level of the solution**

   pH level of the solution also influences the efficiency of the adsorption of water pollutants by the magnetic biochar. There is an optimum value of pH upto which the rate of removal of pollutants by the magnetic biochar increases with increase in the pH and on further increasing the value from the optimum value the removal rate decreases. The type of magnetic biochar and the pollutant present influence this optimum value [52]. For instance, to recover phosphate anion from the aqueous solution with the help of magnesium oxide based magnetic biochar, it was found that at the lower value of pH, the oxides of iron and magnesium get protonated to form $FeOH^+$ and $MgOH^+$ which results in the increase in the adsorption capacity but if the pH is increased further then this results in the deprotonation, as a result of which the surface of the magnetic biochar become negatively charged and repel the phosphate anion and decreasing the adsorption capacity [53].

4. **Effect of adsorbent dosage**

   The dosage of magnetic biochar highly affects the removal of pollutants from water. In general, with increase in the dosage, the rate of removal of effluents also surges but unit adsorption capacity decreases. To elaborate, the adsorption of Cd(II) ions with different doses of iron doped biochar and found that with increase in the dosage of magnetic biochar, the adsorption capacity of pollutants decreases [54].

5. **Contact time**

   The adsorption rate of Cd increases initially and after some period of time, it slows down and become stable because initially large surface area is present and as the substance is adsorbed the adsorption sites on the surface of biochar begins to saturate because of formation of some insoluble substance and this slows down the rate of adsorption as repulsive forces arises between contaminants and in the end, equilibrium is achieved [50].

6. **Initial pollutant concentration**

   A study [55] conducted by Jung et al. found that high initial concentrations of phosphate can enhance the adsorption of phosphate by magnesia ferrite/Biochar. The reason for this enhancement is likely due to the increased contact opportunity between available binding sites on the surface of magnesia ferrite/Biochar and the phosphate in the liquid phase. When the initial concentration of phosphate in the liquid phase is high, there is a greater quantity of phosphate molecules available to come into contact with the binding sites on the surface of the magnesia ferrite/Biochar. This increased availability of phosphate molecules leads to a higher likelihood of successful adsorption onto the binding sites. In other words, a higher initial phosphate concentration provides more opportunities for the binding sites on the magnesia ferrite/Biochar to come into contact with phosphate molecules, resulting in an enhanced adsorption process.

7. **Competitive anions**

Competitive anions can compete with the other pollutant ions and as a result may promote or inhibit a pollutant ion for its adsorption on the surface of the biochar. For instance, acid ions affect the adsorption of heavy metals and organic matter on the surface of magnetic biochar. The order of the effect of the competitive anions on the arsenic adsorption is found as $PO_4^{3-} > SO_4^{2-} > NO_3^-$ [57].

8. **Adsorption temperature**

The adsorption temperature effect the different pollutants differently and also depend on the different biochars nature, i.e., the effect is multilevel and can only be observed by studying different instances:

- It was discovered that as the temperature increased, the biochar coated with Fe_3O_4 became more effective at removing crystal violet and the best temperature for this process was 313 K and also indicate that the process is endothermic [58].
- A study focused on the removal of fluoride from groundwater which utilized both kind of adsorbents, i.e., magnetic and non-magnetic biochar. With increase in adsorption temperature, the removal rate of fluoride decreased gradually. The decrease in removal rate suggests that higher temperatures may have a negative impact on the adsorption capacity of corn straw-derived biochar and magnetic biochar for fluoride [59].

Mechanism of Adsorption of pollutants on magnetic biochar [56]:

Step 1: Ion exchange

This process is used to extract certain ions from the solution and is a reversible exchange reaction.

Step 2: Electrostatic interaction

In this chemical bonds and ionic bonds are formed by the electrostatic attraction and repulsion processes.

Step 3: Surface complexation

In this process, the electron donors and acceptors interact with each other to form various complexes, in which electron pair is donated by the electron donors and metal ions or organic compounds accept these electrons. This process in which ions combine to form new stable ion is called complexation reaction.

Step 4: O-containing functional group and Co-precipitation

The pollutants and the functional groups present on the biochar interact with each other which help in the removal of contaminants and the major role is played by the O-containing functional groups to form linkages. The major functional group present on the surface of biochar is hydroxyl group which form linkage with the heavy metal ion.

In the process of co-precipitation, some co-existing soluble substances are precipitated collectively.

Step 5: *Chemical bond adsorption*

In this two or more atoms present at adjacent sites in a crystal interact strongly with each other to form bonds between them.

Step 6: *Reduction*

Due to strong reduction between magnetic biochar and certain elements, it is feasible to remove these pollutants as shown in Figure 3.11.

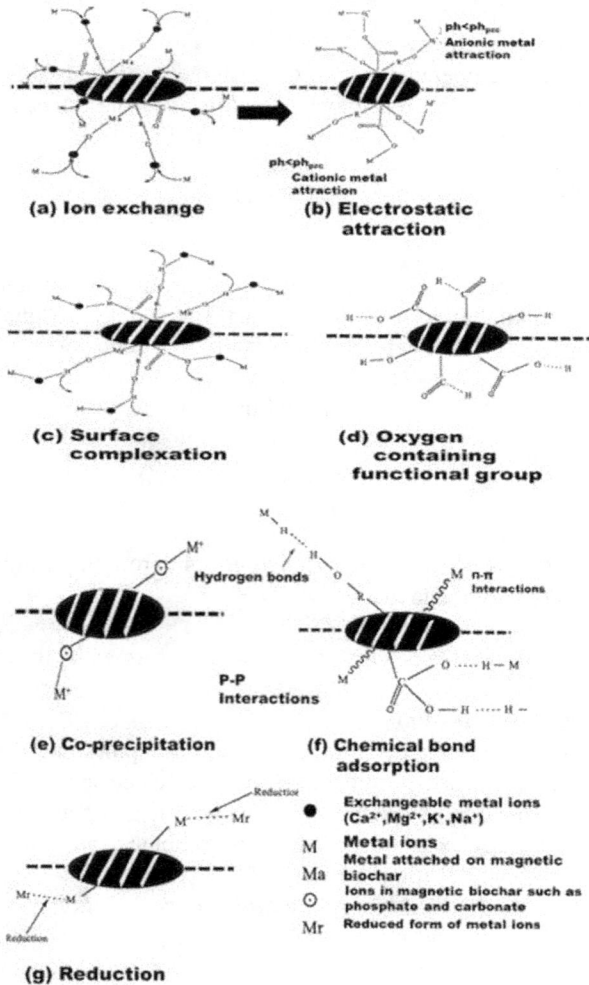

Figure 3.11 Steps of adsorption of pollutants on MBC [56]. (Adapted from Li *et al. Sci. Total Environ.*, 711:134847, 2020. Copyright 2020 Elsevier.)

3.7 CONCLUSION

Calcium-based magnetic biochar, with its combination of calcium modification and magnetic properties, holds significant promise for a range of environmental applications. The incorporation of calcium into biochar enhances its adsorption capacity, cation exchange capacity, and nutrient retention, while the integration of magnetic properties allows for targeted separation and recovery. This unique material offers numerous benefits for environmental remediation, water treatment, soil fertility enhancement, and sustainable agriculture.

In environmental applications, calcium-based magnetic biochar demonstrates effectiveness in removing heavy metals and organic pollutants from water and soil. Its high surface area, modified surface chemistry, and magnetic response enable efficient adsorption and easy separation. The material's magnetic properties facilitate the concentration and targeted removal of contaminants, making it a valuable tool for groundwater remediation.

In the realm of soil and nutrient management, calcium-based magnetic biochar serves as a slow-release fertilizer, providing long-term nutrient supply to plants. The enhanced cation exchange capacity and pH modification properties of biochar contribute to improved soil fertility, nutrient availability, and crop productivity. Additionally, the utilization of calcium-based magnetic biochar promotes sustainable agriculture practices by recycling organic waste materials into a valuable resource.

As with any emerging technology, continued research and development are crucial to further explore the potential of calcium-based magnetic biochar. Optimization of synthesis methods, characterization techniques, and understanding the interactions between calcium, magnetic properties, and the biochar matrix will enhance its performance and expand its applications.

In conclusion, calcium-based magnetic biochar offers a promising solution for environmental challenges, providing efficient and sustainable approaches for water and soil remediation, nutrient management, and agricultural practices. This novel material has the potential to contribute significantly to the development of environmentally friendly technologies and address critical issues in resource management and environmental sustainability.

REFERENCES

1. Q. Zhao, T. Xu, X. Song, S. Nie, S.-E. Choi, and C. Si. Preparation and application in water treatment of magnetic biochar. *Front. Bioeng. Biotechnol.*, **9**:769667, 2021.
2. J. Wu, D. Huang, X. Liu, J. Meng, C. Tang, and J. Xu. Remediation of as (iii) and cd (ii) co-contamination and its mechanism in aqueous systems by a novel calcium-based magnetic biochar. *Journal of hazardous materials*, **348**:10–19, 2018.
3. M. Xu, J. Ma, X.-H. Zhang, G. Yang, L.-L. Long, C. Chen, C. Song, J. Wu, P. Gao, and D.-X. Guan. Biochar-bacteria partnership based on microbially induced calcite precipitation improves cd immobilization and soil function. *Biochar*, **5**(1):20, 2023.
4. P. Wang, W. Chen, R. Zhang, and Y. Xing. Enhanced removal of malachite green using calcium-functionalized magnetic biochar. *International Journal of Environmental Research and Public Health*, **19**(6):3247, 2022.

5. H. Yuan, T. Lu, D. Zhao, H. Huang, K. Noriyuki, and Y. Chen. Influence of temperature on product distribution and biochar properties by municipal sludge pyrolysis. *Journal of Material Cycles and Waste Management*, **15**:357–361, 2013.

6. K. Sun, J. Tang, Y. Gong, and H. Zhang. Characterization of potassium hydroxide (koh) modified hydrochars from different feedstocks for enhanced removal of heavy metals from water. *Environmental Science and Pollution Research*, **22**:16640–16651, 2015.

7. H. Wang and Y. Tang. Research progress of biochar materials for remediation of heavy metal contaminated soil. *Journal of Physics: Conference Series*, **1676**:012081, 2020.

8. J. Wu, Z. Li, D. Huang, X. Liu, C. Tang, S. J. Parikh, and J. Xu. A novel calcium-based magnetic biochar is effective in stabilization of arsenic and cadmium co-contamination in aerobic soils. *Journal of Hazardous Materials*, **387**:122010, 2020.

9. M. Ruthiraan, N. M. Mubarak, R. K. Thines, E. C. Abdullah, J. N. Sahu, N. S. Jayakumar, and P. Ganesan. Comparative kinetic study of functionalized carbon nanotubes and magnetic biochar for removal of Cd^{2+} ions from wastewater. *Korean Journal of Chemical Engineering*, **32**:446–457, 2015.

10. Y. Jiang, J.-L. Gong, G.-M. Zeng, X.-M. Ou, Y.-N. Chang, C.-H. Deng, J. Zhang, H.-Y. Liu, and S.-Y. Huang. Magnetic chitosan–graphene oxide composite for anti-microbial and dye removal applications. *International Journal of Biological Macromolecules*, **82**:702–710, 2016.

11. H. Gao, S. Lv, J. Dou, M. Kong, D. Dai, C. Si, and G. Liu. The efficient adsorption removal of Cr(VI) by using Fe3O4 nanoparticles hybridized with carbonaceous materials. *RSC Advances*, **5**(74):60033–60040, 2015.

12. N. Obinnaya and C. Victor. *Formation and properties of magnetic biochar*, 2022.

13. H. Zeghioud, L. Fryda, H. Djelal, A. Assadi, and A. Kane. A comprehensive review of biochar in removal of organic pollutants from wastewater: Characterization, toxicity, activation/functionalization and influencing treatment factors. *Journal of Water Process Engineering*, **47**:102801, 2022.

14. J. Wu, Z. Li, L. Wang, X. Liu, C. Tang, and J. Xu. A novel calcium-based magnetic biochar reduces the accumulation of As in grains of rice (Oryza sativa L.) in As-contaminated paddy soils. *Journal of Hazardous Materials*, **394**:122507, 2020.

15. O. Das, A. K. Sarmah, and D. Bhattacharyya. A novel approach in organic waste utilization through biochar addition in wood/polypropylene composites. *Waste management*, **38**:132–140, 2015.

16. S. Zhu, X. Huang, D. Wang, L. Wang, and F. Ma. Enhanced hexavalent chromium removal performance and stabilization by magnetic iron nanoparticles assisted biochar in aqueous solution: Mechanisms and application potential. *Chemosphere*, **207**:50–59, 2018.

17. Y. Yi, Z. Huang, B. Lu, J. Xian, E. P. Tsang, W. Cheng, J. Fang, and Z. Fang. Magnetic biochar for environmental remediation: A review. *Bioresource Technology*, **298**:122468, 2020.

18. M. Imran, M. M. Iqbal, J. Iqbal, N. S. Shah, Z. Ul Haq Khan, B. Murtaza, M. Amjad, S. Ali, and M. Rizwan. Synthesis, characterization and application of novel MnO and CuO impregnated biochar composites to sequester arsenic (As) from water: Modeling, thermodynamics and reusability. *Journal of Hazardous Materials*, **401**:123338, 2021.

19. S. Wan, J. Lin, W. Tao, Y. Yang, Y. Li, and F. He. Enhanced fluoride removal from water by nanoporous biochar-supported magnesium oxide. *Industrial & Engineering Chemistry Research*, **58**(23):9988–9996, 2019.

20. Y. Xu, W. Qu, B. Sun, K. Peng, X. Zhang, J. Xu, F. Gao, Y. Yan, and T. Bai. Effects of added calcium-based additives on swine manure derived biochar characteristics and heavy metals immobilization. *Waste Management*, **123**:69–79, 2021.

21. D. R. K. Reddy, S. Anitha, and S. Umadevi. Kantowski-sachs bulk viscous string cosmological model in f(R,T) gravity. *The European Physical Journal Plus*, **129**:1–5, 2014.

22. Z. Feng, R. Yuan, F. Wang, Z. Chen, B. Zhou, and H. Chen. Preparation of magnetic biochar and its application in catalytic degradation of organic pollutants: A review. *Science of the Total Environment*, **765**:142673, 2021.

23. S.-Y. Wang, Y.-K. Tang, C. Chen, J.-T. Wu, Z. Huang, Y.-Y. Mo, K. Zhang, and J.-B. Chen. Regeneration of magnetic biochar derived from eucalyptus leaf residue for lead(II) removal. *Bioresource Technology*, **186**:360–364, 2015.

24. S. Cui, S. Zhang, S. Ge, L. Xiong, and Q. Sun. Green preparation and characterization of size-controlled nanocrystalline cellulose via ultrasonic-assisted enzymatic hydrolysis. *Industrial Crops and Products*, **83**:346–352, 2016.

25. P. Wang, B. Yin, H. Dong, Y. Zhang, Y. Zhang, R. Chen, Z. Yang, C. Huang, and Q. Jiang. Coupling biocompatible Au nanoclusters and cellulose nanofibrils to prepare the antibacterial nanocomposite films. *Frontiers in Bioengineering and Biotechnology*, **8**:986, 2020.

26. P. Zhang, D. O'Connor, Y. Wang, L. Jiang, T. Xia, L. Wang, D. C. W. Tsang, Y. S. Ok, and D. Hou. A green biochar/iron oxide composite for methylene blue removal. *Journal of Hazardous Materials*, **384**:121286, 2020.

27. A. S. Eltaweil, H. Ali Mohamed, E. M. Abd El-Monaem, and G. M. El-Subruiti. Mesoporous magnetic biochar composite for enhanced adsorption of malachite green dye: Characterization, adsorption kinetics, thermodynamics and isotherms. *Advanced Powder Technology*, **31**(3):1253–1263, 2020.

28. X. Yao, L. Ji, J. Guo, S. Ge, W. Lu, L. Cai, Y. Wang, W. Song, and H. Zhang. Magnetic activated biochar nanocomposites derived from wakame and its application in methylene blue adsorption. *Bioresource Technology*, **302**:122842, 2020.

29. Y. Qiu, X. Xu, Z. Xu, J. Liang, Y. Yu, and X. Cao. Contribution of different iron species in the iron-biochar composites to sorption and degradation of two dyes with varying properties. *Chemical Engineering Journal*, **389**:124471, 2020.

30. J. Luo, S. Bo, Y. Qin, Q. An, Z. Xiao, and S. Zhai. Transforming goat manure into surface-loaded cobalt/biochar as PMS activator for highly efficient ciprofloxacin degradation. *Chemical Engineering Journal*, **395**:125063, 2020.

31. S. Tang, S. Tian, C. Zheng, and Z. Zhang. Effect of calcium hydroxide on the pyrolysis behavior of sewage sludge: Reaction characteristics and kinetics. *Energy & Fuels*, **31**(5):5079–5087, 2017.

32. X. Xu, X. Hu, Z. Ding, and Y. Chen. Effects of copyrolysis of sludge with calcium carbonate and calcium hydrogen phosphate on chemical stability of carbon and release of toxic elements in the resultant biochars. *Chemosphere*, **189**:76–85, 2017.

33. S.-Y. Oh, Y.-D. Seo, and K.-S. Ryu. Reductive removal of 2, 4-dinitrotoluene and 2,4-dichlorophenol with zero-valent iron-included biochar. *Bioresource Technology*, **216**:1014–1021, 2016.

34. D. Zhong, Y. Zhang, L. Wang, J. Chen, Y. Jiang, D. C. W. Tsang, Z. Zhao, S. Ren, Z. Liu, and J. C. Crittenden. Mechanistic insights into adsorption and reduction of hexavalent chromium from water using magnetic biochar composite: Key roles of Fe_3O_4 and persistent free radicals. *Environmental Pollution*, **243**:1302–1309, 2018.

35. E. Agrafioti, D. Kalderis, and E. Diamadopoulos. Arsenic and chromium removal from water using biochars derived from rice husk, organic solid wastes and sewage sludge. *Journal of Environmental Management*, **133**:309–314, 2014.

36. C.-D. Dong, C.-W. Chen, T.-B. Nguyen, C. P. Huang, and C.-M. Hung. Degradation of phthalate esters in marine sediments by persulfate over Fe–Ce/biochar composites. *Chemical Engineering Journal*, **384**:123301, 2020.

37. J.-H. Park, J. J. Wang, N. Tafti, and R. D. Delaune. Removal of eriochrome black t by sulfate radical generated from Fe-impregnated biochar/persulfate in fenton-like reaction. *Journal of Industrial and Engineering Chemistry*, **71**:201–209, 2019.

38. J. Yu, L. Tang, Y. Pang, G. Zeng, J. Wang, Y. Deng, Y. Liu, H. Feng, S. Chen, and X. Ren. Magnetic nitrogen-doped sludge-derived biochar catalysts for persulfate activation: Internal electron transfer mechanism. *Chemical Engineering Journal*, **364**:146–159, 2019.

39. H. Fu, S. Ma, P. Zhao, S. Xu, and S. Zhan. Activation of peroxymonosulfate by graphitized hierarchical porous biochar and $MnFe_2O_4$ magnetic nanoarchitecture for organic pollutants degradation: Structure dependence and mechanism. *Chemical Engineering Journal*, **360**:157–170, 2019.

40. D. Huang, Q. Zhang, C. Zhang, R. Wang, R. Deng, H. Luo, T. Li, J. Li, S. Chen, and C. Liu. Mn doped magnetic biochar as persulfate activator for the degradation of tetracycline. *Chemical Engineering Journal*, **391**:123532, 2020.

41. B. Liu, W. Guo, H. Wang, Q. Si, Q. Zhao, H. Luo, and N. Ren. Activation of peroxymonosulfate by cobalt-impregnated biochar for atrazine degradation: The pivotal roles of persistent free radicals and ecotoxicity assessment. *Journal of Hazardous Materials*, **398**:122768, 2020.

42. C. Liu, L. Chen, D. Ding, and T. Cai. From rice straw to magnetically recoverable nitrogen doped biochar: Efficient activation of peroxymonosulfate for the degradation of metolachlor. *Applied Catalysis B: Environmental*, **254**:312–320, 2019.

43. J. Yan, L. Qian, W. Gao, Y. Chen, D. Ouyang, and M. Chen. Enhanced fenton-like degradation of trichloroethylene by hydrogen peroxide activated with nanoscale zero valent iron loaded on biochar. *Scientific Reports*, **7**(1):43051, 2017.

44. Y. Yi, G. Tu, P. E. Tsang, and Z. Fang. Insight into the influence of pyrolysis temperature on fenton-like catalytic performance of magnetic biochar. *Chemical Engineering Journal*, **380**:122518, 2020.

45. B. Li, H. Wang, Y. Lan, Y. Cui, Y. Zhang, Y. Feng, J. Pan, M. Meng, and C. Wu. A controllable floating pDA-PVDF bead for enhanced decomposition of H2O2 and degradation of dyes. *Chemical Engineering Journal*, **385**:123907, 2020.

46. S. Li, Z. Wang, X. Zhao, X. Yang, G. Liang, and X. Xie. Insight into enhanced carbamazepine photodegradation over biochar-based magnetic photocatalyst Fe_3O_4/BiOBr/BC under visible LED light irradiation. *Chemical Engineering Journal*, **360**:600–611, 2019.

47. Y. Zhai, Y. Dai, Jing Guo, Lulu Zhou, Minxing Chen, Hantong Yang, and Liangping Peng. Novel biochar@$CoFe_2O_4$/Ag_3PO_4 photocatalysts for highly efficient degradation of bisphenol a under visible-light irradiation. *Journal of Colloid and Interface Science*, **560**:111–121, 2020.

48. A. Kumar, A. Kumar, G. Sharma, A. H. Al-Muhtaseb, M. Naushad, A. A. Ghfar, C. Guo, and F. J. Stadler. Biochar-templated g-C_3N_4/$Bi_2O_2CO_3$/$CoFe_2O_4$ nano-assembly for

visible and solar assisted photo-degradation of paraquat, nitrophenol reduction and CO_2 conversion. *Chemical Engineering Journal*, **339**:393–310, 2018.

49. S.-F. Jiang, K.-F. Xi, J. Yang, and H. Jiang. Biochar-supported magnetic noble metallic nanoparticles for the fast recovery of excessive reductant during pollutant reduction. *Chemosphere*, **227**:63–71, 2019.

50. M. W. Yap, N. M. Mubarak, J. N. Sahu, and E.C. Abdullah. Microwave induced synthesis of magnetic biochar from agricultural biomass for removal of lead and cadmium from wastewater. *Journal of Industrial and Engineering Chemistry*, **45**:287–295, 2017.

51. H. Li, S. A. Ali Mahyoub, W. Liao, S. Xia, H. Zhao, M. Guo, and P. Ma. Effect of pyrolysis temperature on characteristics and aromatic contaminants adsorption behavior of magnetic biochar derived from pyrolysis oil distillation residue. *Bioresource Technology*, **223**:20–26, 2017.

52. F. Zhang, X. Wang, J. Xionghui, and L. Ma. Efficient arsenate removal by magnetite-modified water hyacinth biochar. *Environmental Pollution*, **216**:575–583, 2016.

53. R. Li, J. J. Wang, B. Zhou, M. K. Awasthi, A. Ali, Z. Zhang, A. H. Lahori, and A. Mahar. Recovery of phosphate from aqueous solution by magnesium oxide decorated magnetic biochar and its potential as phosphate-based fertilizer substitute. *Bioresource Technology*, **215**:209–214, 2016.

54. D. Kołodyńska, J. Bąk, M. Kozioł, and I. Pylypchuk. Investigations of heavy metal ion sorption using nanocomposites of iron-modified biochar. *Nanoscale Research Letters*, **12**:433, 2017.

55. K.-W. Jung, S. Lee, and Y. J. Lee. Synthesis of novel magnesium ferrite ($MgFe_2O_4$)/biochar magnetic composites and its adsorption behavior for phosphate in aqueous solutions. *Bioresource Technology*, **245**:751–759, 2017.

56. X. Li, C. Wang, J. Zhang, J. Liu, B. Liu, and G. Chen. Preparation and application of magnetic biochar in water treatment: A critical review. *Science of the Total Environment*, **711**:134847, 2020.

57. L. Lin, W. Qiu, D. Wang, Q. Huang, Z. Song, and H. W. Chau. Arsenic removal in aqueous solution by a novel Fe-Mn modified biochar composite: Characterization and mechanism. *Ecotoxicology and Environmental Safety*, **144**:514–521, 2017.

58. P. Sun, C. Hui, R. A. Khan, J. Du, Q. Zhang, and Y.-H. Zhao. Efficient removal of crystal violet using $Fe3O4$-coated biochar: The role of the Fe_3O_4 nanoparticles and modeling study their adsorption behavior. *Scientific Reports*, **5**(1):1–12, 2015.

59. D. Mohan, S. Kumar, and A. Srivastava. Fluoride removal from ground water using magnetic and nonmagnetic corn stover biochars. *Ecological Engineering*, **73**:798–808, 2014.

4 Calcium-Based Metal-Organic Frameworks

Simranpreet Kaur and Sanjeev Gautam
Advanced Functional Materials Laboratory, Dr. S.S. Bhatnagar
University Institute of Chemical Engineering and Technology
Panjab University, Chandigarh, India

4.1 INTRODUCTION

Metal–organic frameworks (MOFs), also referred to as porous coordination polymers or PCPs, belong to a developing category of porous materials. These materials are created by combining metal-containing nodes (also called secondary building units or SBUs) with organic linkers [1]. They are composed of metal-containing nodes or clusters, known as secondary building units (SBUs), which are interconnected by organic linkers. The resulting 3D network structure of MOFs combines the robustness of inorganic components with the versatility and diversity of organic ligands [2]. The design and synthesis of MOFs offer an exceptional level of control over their structure and properties. By carefully selecting the metal nodes and organic linkers, researchers can create a virtually limitless array of MOF architectures with distinct pore sizes, shapes, and functionalities. This tunability allows MOFs to be tailored for specific applications and desired properties, making them highly versatile materials [3].

One of the most remarkable features of MOFs is their exceptional porosity. The pores and cavities within MOFs can exhibit a vast internal surface area, sometimes reaching thousands of square meters per gram of material. This large surface area, combined with the precisely designed pore sizes, enables MOFs to efficiently adsorb, sense, and store gases, liquids, and even biomolecules. The remarkable gas storage capabilities of MOFs have attracted considerable attention for applications such as carbon capture, hydrogen storage, and gas separation [4–6].

In addition to gas storage and catalysis, MOFs have found applications in drug delivery systems. The porous structure of MOFs allows for the encapsulation and controlled release of therapeutic agents, offering advantages such as improved drug stability, targeted delivery, and sustained release profiles. The ability to modify the surface chemistry of MOFs further enables their integration with biological systems for various biomedical applications [7–9].

DOI: 10.1201/9781003360599-4

The field of MOFs has witnessed explosive growth over the years. The number of MOF structures reported in the scientific literature has surged, reflecting the increasing interest and research efforts in this field. Moreover, the applications of MOFs have extended beyond traditional areas, leading to collaborations between chemists, physicists, engineers, and biologists, fostering interdisciplinary research, and innovation. Amongst the diverse range of MOFs, calcium-based metal-organic frameworks (Ca-MOFs) have garnered significant attention due to their distinct properties and potential applications. Ca-MOFs are constructed by linking calcium-containing nodes or clusters with organic ligands, resulting in a three-dimensional porous structure. This introduction aims to provide an overview of Ca-MOFs, highlighting their importance, synthesis, and notable characteristics [10].

Ca-MOFs offer several key advantages that contribute to their significance. First, the abundance and economic viability of calcium make Ca-MOFs more accessible compared to those based on rare or expensive metals. Additionally, calcium's biocompatibility opens up opportunities for biomedical applications, including drug delivery systems [11]. This combination of abundance and biocompatibility makes Ca-MOFs highly attractive for various research areas.

The tunability of Ca-MOFs is another defining characteristic that adds to their importance. By selecting specific organic linkers and adjusting synthesis conditions, the structure, pore size, and functionality of Ca-MOFs can be precisely tailored for various applications. This versatility allows for customization and optimization for specific purposes, ranging from gas storage and separation to catalysis and biomedical applications [12].

The high porosity of Ca-MOFs is another significant feature that contributes to their importance. The porous structure provides a large internal surface area, enabling efficient adsorption and storage of gases, liquids, and even biomolecules. This property has applications in carbon capture, hydrogen storage, and gas separation, as well as controlled drug delivery systems and other biomedical applications [13].

Furthermore, Ca-MOFs exhibit catalytic activities due to the presence of calcium ions and the incorporation of functional groups into the framework [14]. These catalytic properties enable Ca-MOFs to facilitate various chemical transformations, including selective organic reactions and photocatalysis.

As a rapidly evolving field, research on Ca-MOFs continues to advance, exploring new synthetic strategies, characterizing their structures and properties, and investigating their applications in diverse areas. The versatility, biocompatibility, tunability, and unique properties of Ca-MOFs make them promising materials in the realm of MOFs, with potential implications in energy, environment, healthcare, and other scientific and technological domains [15].

This chapter delves into the synthesis, characterization, properties, and applications of Ca-MOFs. It explores various synthetic methods, structural characterization techniques, properties, and functionalities, as well as applications in gas storage, catalysis, and biomedicine. Additionally, recent advances, challenges, and future directions in Ca-MOF research are examined. Through this comprehensive exploration, we aim to provide a deeper understanding of the importance and potential of Ca-MOFs as versatile materials in the realm of MOF.

4.2   SYNTHESIS METHODS AND ITS EFFECT ON THE STRUCTURE OF Ca-MOF

Crafting the structure, properties, and potential applications of Ca-MOFs hinges on their synthesis methods. The choice of synthetic techniques notably shapes the morphology, crystalline quality, porosity, and overall functionality of resultant MOFs. This chapter conducts a comprehensive investigation into diverse synthesis approaches applied to Ca-MOFs, elucidating their impact on material attributes and shedding light on the multifaceted realm of these captivating porous substances.

4.2.1   PRECURSORS AND BUILDING BLOCKS FOR Ca-MOF AND THEIR SIGNIFICANCE

The synthesis of Ca-MOFs relies on carefully selected precursors and building blocks, which are fundamental in shaping the resulting structure, properties, and potential applications of Ca-MOFs. These components hold immense significance in determining the unique characteristics and versatility of Ca-MOFs.

Ca-MOFs, a subtype of MOFs with calcium as its metal centre, are Ca-MOFs. Contrary to MOFs made of transition and post-transition metals (such as Zr, Fe, Co, Ni, Cu, Zn, etc.), which have a propensity to form frequently observed SBUs and topology and thus make it easier to design and implement targeted structure and functionality, it is much more difficult to predict the coordination geometry and structural topology of Ca-MOFs. This may be explained by the fact that the bonding interactions between calcium and organic ligands (typically carboxylates or phosphates) are more ionic, and as a result, the coordination mode of calcium and the topology of Ca-MOFs heavily depend on the makeup of the organic ligands and the synthetic conditions.

Calcium salts, especially calcium chloride ($CaCl_2$) or calcium nitrate ($Ca(NO_3)_2$), are frequently utilised as the calcium source in Ca-MOF production. The degree of solubility as well as reaction during the process of synthesis are influenced by the calcium salt used, and this can have an effect on the crystallisation along with purity of the resultant Ca-MOF. Calcium salts are useful for large-scale production and prospective biological uses of Ca-MOFs since they are easily accessible, inexpensive, and frequently thought to be non-toxic. Organic linkers are crucial ingredients in the creation of Ca-MOFs because they link calcium ions to create the framework structure. As linkers, a variety of organic ligands can be used, including carboxylates (such as terephthalic acid and benzenedicarboxylate), imidazoles, pyridines, and other nitrogen-containing substances [16, 17]. The most frequent organic ligands added to MOFs are carboxylates, particularly multi-carboxylate ligands. Pure carboxylates fall under this category, as do multifunctional compounds having both pure carboxylates and additional functional groups like hydroxyl, imidazole, triazole, tetrazole moieties, etc. Phosphate ligands are only found in a relatively small number of transition metal-based MOFs, although they are often present in Ca-MOFs. This may be partially attributable to the high charge density and ionic makeup of Ca^{2+}, which cause it to strongly connect with phosphates. Phenols, thiols, sulphates,

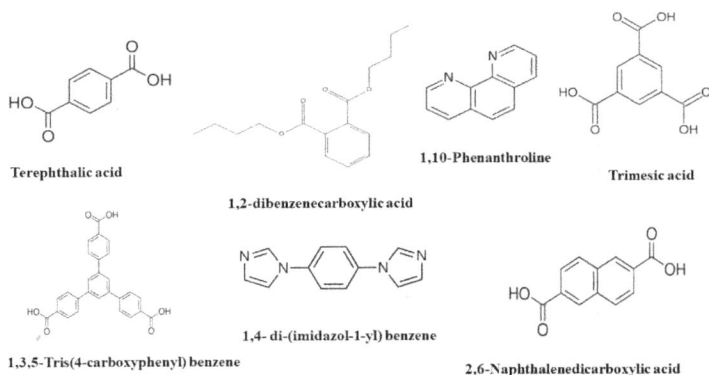

Figure 4.1 Various ligands that are commonly used in the synthesis of Ca-based metal-organic frameworks.

imidazoles, and other substances fall under third group. The stability of the resultant Ca-MOF, the coordination geometry surrounding the calcium ions, and the material's functioning are all influenced by the organic linker used. As a result, the applications of Ca-MOFs in domains including catalysis, sensing, and drug delivery have increased. Functionalized organic linkers can impart particular characteristics or reactivity to Ca-MOFs [10], some of the common ligands used are shown in Figure 4.1.

In order to create Ca-MOF, solvents are essential since they function as a reaction's medium and have an impact on the crystal development and shape of the finished product. Water, other organic solvents (such as methanol, ethanol, dimethyl-formamide), and solvent combinations are frequently utilised as solvents in the manufacture of Ca-MOFs [18]. The precursor solubility, reaction kinetics, and creation of certain Ca-MOF phases are all impacted by the solvent choice, which enables control over crystallinity, size, and characteristics. To provide further functionality or regulate the crystallisation process, co-ligands or modulators can be introduced into the Ca-MOF synthesis. In order to provide Ca-MOFs different characteristics or reactivity, co-ligands are secondary organic ligands that are combined with the primary linker. Enhancing control over the final Ca-MOF structure is possible by using modulators, such as organic bases or acids, to alter the pH, rate of hydrolysis, and crystal formation throughout the synthesis process.

4.2.2 SYNTHESIS METHODS

Method of synthesis of MOFs significantly effect their surface topology, morphology surface area and porosity which directly impact the application of MOFs. Table 4.1 shows some CaMOFs along with their formulas, method of synthesis and applications.

Table 4.1
Various Ca-MOFs Synthesis Methods and Their Applications

MOF	Formula	Synthesis	Application	References
Ca (square)	Ca(SA).3H$_2$O	Hydrothermal method	Hydrocarbons separation	[19]
CYCU-1	[Ca(SDB)]·0.45H$_2$O	Microwave-assisted reaction	N$_2$, H$_2$, CO$_2$ adsorption	[20]
CaBDC	CaBDC.3H$_2$O	Ion exchange method	Na ion batteries	[21]
CaBTC	Ca$_3$(BTC)$_2$	Ion exchange method	Hydrogen storage	[16]
CaHBTC	Ca(HBTC)	One-pot self-assembly reaction	Hydrogen storage-assembly reaction	[16]
CaBDC	CaBDC	Microwave synthesis	pH responsive drug delivery	[22]

4.2.2.1 Solvothermal and Hydrothermal Method

Solvothermal and hydrothermal methods involve the use of high-temperature and high-pressure conditions in an autoclave or pressure vessel as shown in Figure 4.1. Ca-MOF crystal formation is aided by heating the reaction mixture, which contains organic linkers and calcium salts. While water is the solvent in hydrothermal synthesis, organic solvents are frequently used in solvothermal synthesis as the reaction media (Figure 4.2). The well-defined Ca-MOF structures produced by these techniques have a high crystallinity and can be produced by controlled crystal development. A MOF, (H$_3$O$^+$)$_2$[Ca(NDC)(C$_2$H$_5$O)(OH)]$_{41}$.1H$_2$O, was produced via the solvothermal reaction of a combination of calcium acetylacetonate and 1,4-naphthalenedicarboxylic acid (H$_2$NDC) in an ethanol and distilled water solution. This MOF has a distinct fluorescence feature, having a fluorescence peak at 395 nm (ex = 350 nm) at room temperature that is blue-shifted in comparison to the free

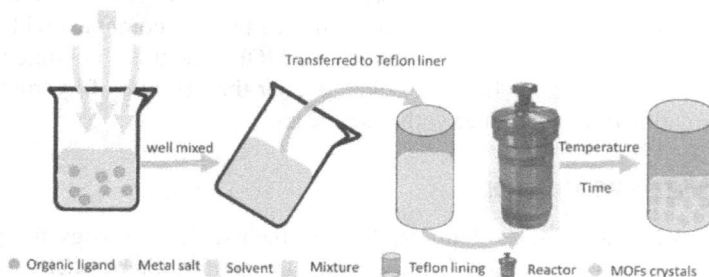

Figure 4.2 Solvothermal/hydrothermal method of synthesis [24]. (Adapted with permission under OACC from Bian *et al. Processes*, 6(8):122, 2018. Copyright 2018 MDPI.)

H_2NDC ligand. It also has a novel structure made up of calcium clusters and H_2NDC linker anions [23].

4.2.2.2 Microwave-Assisted Synthesis

In microwave-assisted synthesis, the reaction mixture is exposed to microwave radiation, which speeds up the reaction kinetics and improves crystallization. Comparing the reaction periods of this method to more conventional solvothermal or hydrothermal techniques, it provides quick and effective Ca-MOF synthesis. It is possible to precisely regulate the reaction temperature with microwave irradiation, and it can result in the development of distinctive Ca-MOF structures. The quick synthesis of nanopore materials in hydrothermal environments and the utilisation of organic synthesis have both benefited greatly by the widespread usage of microwave coherence methods. In this procedure, microwave radiation (MW) is employed to obtain the energy needed for the reaction. The tiny and homogenous particle size distribution, superb morphological control, and quick crystallisation process are only a few benefits of this method. Cr-MIL-10 is the first MOF created using this technique. This approach of using a microwave oven has also been used to synthesise MOF, which comprises Fe^{3+}, Al^{3+}, Cr^{3+}, V^{3+}, and Ce^{3+} [25].

4.2.2.3 Ionothermal Synthesis

Ionothermal synthesis represents a specialized and distinctive approach for creating Metal-Organic Frameworks (MOFs), the ionic liquid serves a dual purpose as both the solvent and reaction medium, effectively facilitating the formation of MOF crystals by solvating the metal ions and ligands. The distinctive properties of the ionic liquid, such as its low vapor pressure and high thermal stability, establish a specialized reaction environment that significantly influences the formation and morphology of the resulting MOF crystals. Through ionothermal synthesis, the precisely tailored ionic liquid conditions yield MOFs with unique structures and properties, making them well-suited for targeted applications, such as gas storage, catalysis, and separation processes. Furthermore, this method presents an environmentally favorable alternative to traditional solvent-based approaches, as ionic liquids are often considered safer and more sustainable. However, the careful selection of appropriate ionic liquids and optimization of reaction conditions are essential to achieve the desired MOF structures and properties successfully [26].

4.2.2.4 Mechanochemical Synthesis

Mechanochemical synthesis relies on mechanical force, such as ball milling or grinding, to induce chemical reactions between the precursors. The reaction mixture, containing calcium salts and organic linkers, is subjected to mechanical force in a ball mill or other milling equipment. The mechanical force promotes bond breaking and reformation, resulting in the formation of Ca-MOF crystals. This method offers a solvent-free and energy-efficient approach to Ca-MOF synthesis, reducing the environmental impact and simplifying the purification process. Mechanochemical synthesis can lead to the formation of metastable phases and enhance the reactivity of the precursors, resulting in unique Ca-MOF structures. The study by Crickmore et al. demonstrates a sustainable approach for synthesizing Ca-MOFs using waste materials and recyclable plastics. The solvent-minimized mechanochemical synthesis involving eggshells and linkers from PET bottles resulted in the successful preparation of $Ca(BDC)(H_2O)_3$, highlighting the eco-friendly nature of the method [27].

4.3 CHARACTERISATION

The combination of these structural characterization techniques provides a comprehensive understanding of the composition, crystallinity, morphology, and properties of Ca-MOFs. These analyses are essential for confirming the successful synthesis of Ca-MOFs, verifying their structural integrity, and tailoring their properties for specific applications in fields such as gas storage, catalysis, and biomedical research.

4.3.1 X-RAY DIFFRACTION (XRD)

X-ray diffraction (XRD) is a powerful technique extensively used to determine the crystal structure and phase purity of Ca-MOFs. By directing X-rays at the sample, researchers obtain a diffraction pattern that reveals valuable insights into the periodic arrangement of atoms within the crystal lattice, it is a crucial technique used in the analysis of Ca-MOFs. It plays a significant role in determining the crystal structure and confirming the purity of the materials. By studying the diffraction pattern, researchers can gain valuable insights into the periodic arrangement of atoms within the MOF's lattice. XRD is essential for verifying the structural integrity during synthesis, ensuring consistency and reliability in research. The crystal structure information obtained through XRD is vital for understanding properties such as pore size, surface area, and active sites, all of which are essential for various applications, including gas storage, catalysis, and drug delivery. In the schematic crystal structure of CaBDC·3H$_2$O, it is evident that the compound possesses a layered structure with coordinated water molecules sharing the edges of the polyhedra. Figure 4.3 shows the powder X-ray diffraction pattern of the samples prepared at 2h degree from 5° to 60°. Although the pattern of CaBDC has not been fully analyzed, the obtained CaBDC.3H$_2$O pattern exhibits sharp and narrow peaks, which closely match with those of Calcium terephthalate trihydrate [21].

4.3.2 SCANNING ELECTRON MICROSCOPY (SEM)

Scanning Electron Microscopy (SEM) is a valuable characterization technique used in the study of MOFs. It allows researchers to investigate the surface morphology, particle size, and structural features of MOF materials at high resolution. SEM plays a crucial role in understanding the physical properties and microstructure of MOFs, providing valuable information for their synthesis, modification, and various applications. In SEM, a focused beam of electrons scans the surface of the MOF sam-

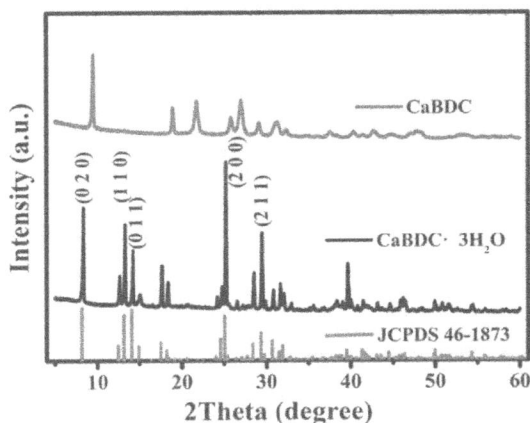

Figure 4.3 X-ray diffraction pattern of the CaBDC·3H₂O possesses a layered structure with coordinated water molecules sharing the edges of the polyhedra [21]. (Adapted with permission from Xiao *et al.*, *Mater. Lett.* 286:129264, 2021. Copyright 2021 Elsevier).

ple. When the high-energy electrons interact with the atoms in the MOF's crystal structure, several types of signals are generated and detected to create detailed images. SEM provides high-resolution imaging of MOF surfaces, allowing researchers to visualize the microstructure, surface features, and texture of MOF crystals at a nanoscale level. This capability is essential for understanding the morphology and shape of MOF particles, which directly impact their properties and performance. SEM offers detailed information about the surface topography of MOFs, by revealing the surface characteristics, such as surface roughness, pores, and particle size distribution, researchers can gain insights into the surface properties and the accessibility of active sites within MOF crystals. Calcium-based MOFs have been investigated as catalysts for esterification and transesterification reactions in biodiesel production from waste cooking oil (WCO). SEM images of Ca-MOF are presented in Figure 4.4(a,b) at distinct magnifications: 10 and 5 μm. The images reveal that the Ca-MOF particles exhibit a cubical shape, with an estimated size ranging from 5 to 7 μm [28].

4.3.3 THERMOGRAVIMETRIC ANALYSIS (TGA)

Thermogravimetric Analysis (TGA) is a powerful thermal analysis technique used to study the thermal stability and decomposition behavior of materials. In TGA, the weight of a sample is continuously monitored as the temperature is gradually increased under controlled conditions. The resulting thermogravimetric curve provides valuable information about the sample's weight loss or gain as a function of temperature or time. The fundamental principle of TGA is based on the fact that most substances undergo thermal decomposition, sublimation, desorption, or oxidation/reduction reactions when subjected to a controlled increase in temperature.

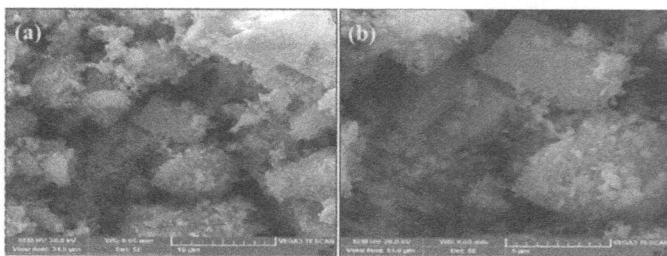

Figure 4.4 Scanning Electron Microscope images of Ca-MOF investigated as catalysts for esterification and transesterification reactions in biodiesel production exhibit a cubical shape with size ranging from 5 to 7 μm [28]. (Adapted with permission from Jamil *et al. Energy Convers. Manag,* 215:112934, 2020. Copyright 2020 Elsevier.)

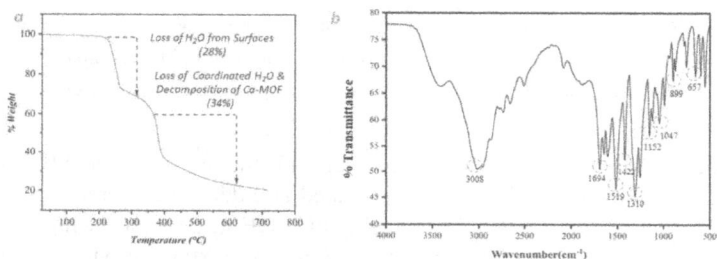

Figure 4.5 (a) Thermogravimetric Analysis of Ca-MOF synthesised by hydrothermal method, and (b) FT-IR spectra of MOF showing all the stretching and vibration bands [29]. (Adapted with permission under OACC from Jha *et al. Mater. Adv,* 4:3330–3343, 2023. Copyright 2023 RSC.)

During the TGA experiment, the sample is placed in a crucible or pan, and the crucible is heated at a constant rate while the weight of the sample is continuously recorded After sample activation, a thermogravimetric device was used to examine the thermal stability of the synthesised Ca-MOF (Figure 4.5(a)). The elimination of free water molecules from the MOF surfaces occurs between 220 and 267 °C, according to the TGA data. At temperatures between 267 and 393 °C, a second mass loss of up to 34% occurs, which is attributable to the partial breakdown of the organic ligand and the evaporation of bound water molecules. After reaching this temperature, the MOF structure collapses [29].

4.3.4 FOURIER TRANSFORM INFRARED SPECTROSCOPY (FT-IR)

Fourier Transform Infrared Spectroscopy (FT-IR) is a valuable analytical technique in the analysis of MOFs due to its ability to provide critical information about the chemical composition, bonding, and functional groups present in these materials. FT-IR allows researchers to identify the various functional groups present in MOFs. The absorption peaks in the FT-IR spectrum correspond to specific vibrational modes of the functional groups, providing valuable information about the MOF's organic linkers, ligands, and coordinating groups. FT-IR spectra offer insights into the coordination bonds and interactions between metal ions and organic ligands in the MOF framework. It helps in understanding the chemical structure and connectivity of the MOF building blocks. FT-IR can be employed to quantify guest molecules or solvents within the MOF structure. By measuring the intensity of specific absorption peaks, researchers can estimate the amount of adsorbed species or the degree of solvent inclusion. Changes in the FT-IR spectrum upon exposure to different gases or solutions can provide insights into gas sorption, catalytic activity, and interactions with guest molecules. The FT-IR spectrum of the synthesized Ca-MOF is shown in Figure 4.5(b). Peaks observed at 3308, 1152, and 1047 cm^{-1} indicate the presence of O–H stretching through hydrogen bonding with water and CaO, while the bending mode of the undissociated water molecule is detected at 1694 cm^{-1}. The peaks at 1519 cm^{-1} correspond to asymmetric O–C–O stretch, and those at 1422 and 1310 cm^{-1} correspond to symmetric O–C–O stretch. The sharp peaks observed at 1694 and 1422 cm^{-1} indicate an O–C–O stretch with a bidentate bridge caused by an ionic carbonyl bond. Additionally, peaks at 657 and 899 cm^{-1} suggest the presence of Ca–O stretching as a component of the calcium oxo cluster [29].

4.4 FEATURES OF Ca-MOF

Ca-MOFs possess several notable features that make them unique and attractive for various applications. Its major features are shown in Figure 4.6.

4.4.1 ABUNDANCE OF CALCIUM

Calcium (Ca) is the fifth most abundant element in the Earth's crust, with an estimated abundance of approximately 3.63% by mass. It is a vital alkaline earth metal and plays a crucial role in various natural processes, as well as in human health and industry. It regulates functions in skeletal muscles, nerve conduction, and fibrin polymerization. In protozoa, plants, and animals, calcium serves as a critical intracellular messenger. Calcium-transporting systems in the plasma membrane and organelles maintain the ionic concentration of calcium in different cellular compartments, responding to physiological demands [30]. Calcium is an essential mineral for human health, playing a critical role in bone and teeth formation, nerve transmission, muscle function, blood clotting, and other physiological processes. It is obtained through dietary sources, such as dairy products, leafy green vegetables, nuts, and fortified foods [31]. Calcium is commercially extracted from natural sources, such as limestone, for various industrial applications. It is used in the production of cement, lime, calcium carbide, and other important materials. The abundance of calcium makes it a widely accessible and economically favorable element for various industrial, agricultural, and biological applications. Its widespread availability and importance in numerous natural processes and human health make it a fundamental element with diverse implications across different disciplines [32].

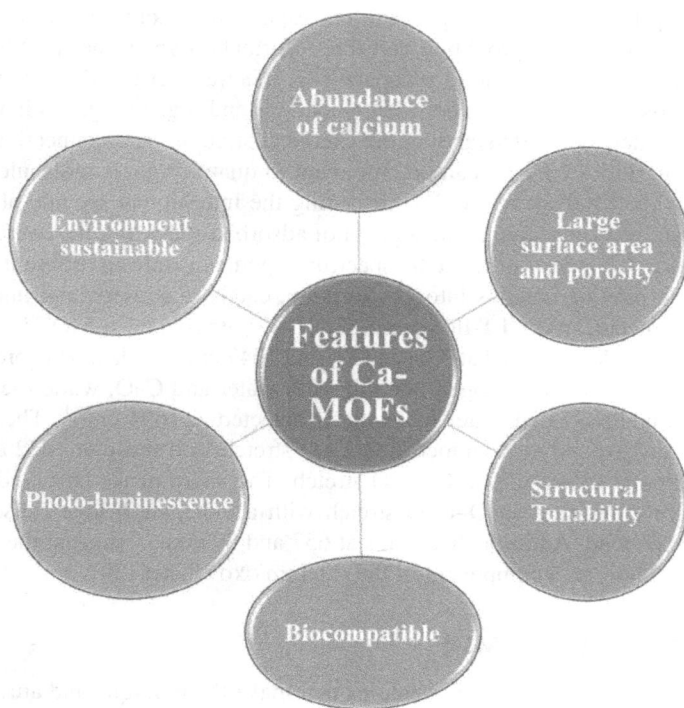

Figure 4.6   Some typical features of Ca based metal-organic frameworks.

4.4.2   LARGE SURFACE AREA AND HIGH POROSITY

Ca-MOFs have a porous three-dimensional structure that offers a sizable interior surface area. Because of the high porosity, gases, liquids, and even biomolecules may be stored and adsorb effectively. The increased surface area permits improved gas adsorption capabilities, making Ca-MOFs ideal materials for carbon capture, gas storage, and separation operations. $[Ca_2Na_2(L)_2(H_2O)_6]nnH_2O$, which has a high BET surface area of 1419 m^2g^{-1} and a large pore volume of 0.74 cm^3g^1, was produced by Saha et al. using calcium and chelidamic acid (H3L) with open metal sites. The CO_2 adsorption capacity of this MOF is astoundingly large, reaching 3.4 mmol g^{-1} at 298 K and 1 bar, with a heat of adsorption of 35 kJ mol^{-1}. The high CO_2 capacity as being due to of the material's huge surface area and robust quadruple interaction with the exposed metal sites. This MOF is an excellent choice for CO_2 collection from flue gas since it can still adsorb 1.61 mmol g^{-1} CO_2 at 0.15 pressure [33].

4.4.3   STRUCTURAL TUNABILITY

Calcium-based Metal-Organic Frameworks (Ca-MOFs) exhibit tunable structural diversity due to the flexibility in designing organic linkers and metal coordination geometries. By selecting different organic linkers and metal ions, a wide range of MOF architectures can be systematically tailored. This versatility allows the creation of various structures, including one-dimensional chains, two-dimensional layers, and three-dimensional frameworks. Additionally, post-synthetic modifications and incorporation of mixed-metal ions further enhance the tunability. The ability to control pore size, surface functionality, and dynamic behavior, such as breathing phenomena, makes Ca-MOFs promising materials for applications in gas storage, catalysis, drug delivery, and environmental remediation, among others [10].

4.4.4 BIOCOMPATIBILITY

Calcium-based Metal-Organic Frameworks (Ca-MOFs) demonstrate excellent biocompatibility, making them promising for various biomedical applications. Their low toxicity and non-immunogenic nature ensure they can coexist with living tissues without causing adverse reactions. Some Ca-MOFs are biodegradable, breaking down harmlessly in the body. They maintain stability in biological environments and have been found to be compatible with different cell types. Ca-MOFs can serve as drug delivery carriers, releasing therapeutic agents safely, enhancing drug stability and reducing side effects [34]. Additionally, they can be functionalized for biomedical imaging and encapsulate biomolecules without compromising functionality. These characteristics open doors to applications in tissue engineering, regenerative medicine, drug delivery, and medical imaging, holding great potential for advancing biomedical research and healthcare. Thorough assessments and studies are vital to ensure their safe and effective utilization in medical settings [35].

4.4.5 ENVIRONMENTALLY SUSTAINABLE

An environmentally sustainable Ca-MOF is designed, synthesized, and utilized with a strong emphasis on reducing its environmental impact and promoting long-term sustainability. Green synthesis methods, involving benign solvents, minimal waste generation, and low energy consumption, are employed to ensure eco-friendliness. Sustainable Ca-MOFs are crafted from renewable resources, minimizing reliance on scarce or non-renewable materials. Biodegradable MOFs are designed to break down harmlessly, reducing environmental burden [36]. Their low toxicity ensures safe coexistence with living organisms and ecosystems. Recyclability allows for reduced waste and resource conservation. These MOFs find applications in environmental remediation, capturing and removing pollutants, as well as carbon capture and storage (CCS) for mitigating climate change. By focusing on long-term stability and responsible material utilization, environmentally sustainable Ca-MOFs contribute to greener technologies and sustainable development, making them essential components of environmentally conscious solutions for various environmental and industrial challenges [37].

4.4.6 PHOTOLUMINESCENCE (PL)

Photoluminescence (PL) is a crucial property of many Metal-Organic Frameworks (MOFs), and their emission of light can have various origins, such as ligand-based, metal-based, guest-induced, ligand-to-metal charge transfer (LMCT), metal-to-ligand charge transfer (MLCT), ligand-to-ligand charge transfer (LLCT), or combinations of these phenomena. Luminescent MOFs (LMOFs) show great promise as probing materials in chemical sensing and detection, as well as phosphors for lighting-related applications. Extensive research has been conducted on LMOFs and their diverse applications, including several calcium-based LMOFs.

Jana et al. introduced a bimetallic Luminescent Metal-Organic Framework (LMOF) that demonstrates sensitive and selective detection of nitroaromatics. The MOF, [{Ca$_6$(H$_2$O)18][Ag$_6$(MNA)$_6$]}·[Ag$_6$(HMNA)$_6$]·20H$_2$O] (H$_2$MNA = 2-mercapto nicotinic acid), was synthesized using a diffusion method. The emissions from intraligand, cluster-based LMCT, and metal-centered transitions of the Ag6 core result in three emission bands around 450, 490, and 530 nm under an excitation wavelength of 375 nm. Upon exposure to nitroaromatics such as picric acid (PA), p-nitrophenol (PNP), and 2-nitrotoluene (2-NT), the emission intensity significantly decreased, with quenching constants (Ksv) ranging from 39, 104–118. The rigid nature of the framework and the high HOMO of [Ag$_6$(H$_2$MNA)$_6$] cluster facilitated efficient electron transfer, contributing to the high quenching efficiency. Particularly, the quenching efficiency with 2-NT could be attributed to dipole-dipole interactions between the MOF and 2-NT, which enhance the electron transfer process and further increase the quenching efficiency. This LMOF shows potential as a sensitive probe for detecting nitroaromatic compounds in various applications [38].

4.5 APPLICATIONS

Ca-MOFs have a wide range of applications due to their tunable structures and properties. Some of the key applications of Ca-MOFs are given below.

4.5.1 GAS STORAGE AND SEPARATION

Ca-MOFs also contribute significantly to the capture and containment of harmful gases, particularly within the context of mitigating greenhouse gas effects. Their customizable characteristics position them as strong contenders for specific gas adsorption, thus aiding in emission reduction and tackling climate change concerns. MOFs, distinguished by their unparalleled structural versatility, finely adjustable porosity, and adaptable surface activity, exhibit substantial promise in adsorption and separation applications. While research on Ca-MOFs in this domain lags behind those constructed with transition metals, they boast distinct advantages such as abundant Ca^{2+} content, low weight, and robust thermal stability. In recent years, the scrutiny of Ca-MOFs for gas adsorption and separation capabilities has expanded, encompassing applications like CO$_2$ capture, noble gas and hydrocarbon separation, as well as the capture of metal ions and diverse molecules. Ca-5TIA-MOF is an early example of a Ca-MOF studied for its CO_2 adsorption properties. Its structure is composed of 5-(1,2,4-triazoleyl) isophthalic acid ($H_2$5TIA) and Ca^{2+} ions, forming a porous framework with a pore size of 3.6 Å. This pore size is smaller than the kinetic diameter of N_2 (3.64 Å) but larger than that of CO_2 (3.3 Å). Gas adsorption experiments revealed a CO_2 uptake of 1.12 mmol g^{-1} on Ca-5TIA-MOF, which is comparable to other reported Ca-MOFs. Interestingly, when water is absent during synthesis, a gel-phase material called Ca-5TIA-Gel is formed instead of crystalline Ca-5TIA-MOF. Ca-5TIA-Gel demonstrated a 20% increase in CO_2 adsorption capacity [39].

The separation and purification of light hydrocarbon mixtures into pure species is a challenging yet crucial industrial process. While the investigation of Ca-MOFs for differentiating hydrocarbons is limited, they show significant potential for this application. Recently, Chen et al. reported an ultra-microporous MOF called $Ca(C_4O_4)(H_2O)$, known as UTSA-280, which efficiently separates C_2H_4 and C_2H_6 through molecular sieving. UTSA-280 possesses rigid 1D channels with a cross-sectional area of approximately 14.4 $Å^2$, enabling the diffusion of ethylene (cross-sectional area 13.7 $Å^2$) while excluding ethane (cross-sectional area 15.5 $Å^2$). Multicomponent breakthrough experiments further confirmed its efficiency in separating ethylene/ethane with high ethylene productivity under ambient conditions. The isosteric heat for ethylene adsorption ranged from 20.5 to 35.0 kJ mol^{-1}, lower than MOFs with open metal sites (OMSs), indicating potential practicality in regenerating this MOF under mild conditions [40].

4.5.2 CATALYSIS AND PHOTOCATALYSIS

The presence of calcium ions within Ca-MOFs confers notable Lewis acidity, rendering them potential catalysts for specific reactions [41]. Calcium, being a cost-effective and widely available metal, poses minimal environmental concerns due to its inherent safety and abundance in the natural environment. This has led to the development of numerous Ca-MOFs exhibiting robust catalytic capabilities in various processes. One of the studies in which two halide salt ligands, 1,3-bis(carboxymethyl)imidazolium bromide (BCMIM·HBr) and BCMIM·HCl, were combined with a calcium salt to create two corresponding MOFs known as bcmim-Ca_1 and bcmim-Ca_2, respectively. Pastor et al. investigated the catalytic properties of these MOFs in the Friedländer reaction of 2-aminobenzaldehydes. Using a 10 mol% bcmim-Ca1 catalyst, they achieved full conversion after 90 minutes at 80 °C in the presence of 10 equiv. of pentane-2,4-dione. The quinoline product was isolated with a yield greater than 99%, showcasing the potential of this catalyst for the reaction. This study also revealed that subtle differences in the MOF structures, such as the counter ions of the organic linkers, can influence the catalytic activity of the heterogeneous catalyst. Additionally, the catalytic systems were effective for both 2-aminobenzophenones and 2-aminobenzaldehydes, demonstrating a new application of MOFs for quinoline synthesis [42].

The utilization of Ca-MOFs as photocatalysts has emerged as a promising avenue in the field of catalysis and energy conversion. These materials exhibit unique properties that make them suitable candidates for driving photochemical reactions. The

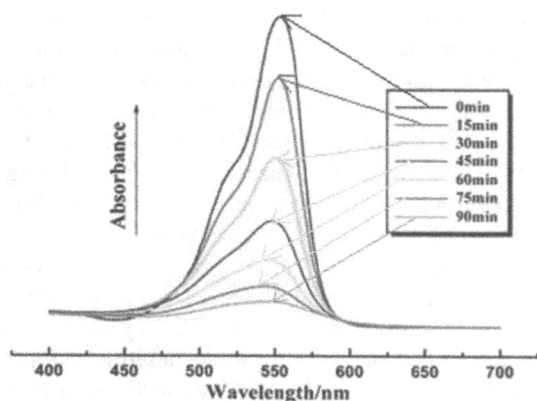

Figure 4.7 The adsorption spectra of the $Ca_2[SiMo_{12}O_{40}] \cdot nH_2O$ MOF catalyzed degradation of Rhodamine-B (RhB) under UV irradiation [43]. (Adapted with permission from Hao *et al. Inorg. Chem. Commun.*, 41:19–24, 2014. Copyright 2014 Elsevier.)

inherent structural diversity and tunable pore sizes of Ca-MOFs can facilitate the accommodation of various reactants and intermediates, enhancing their photocatalytic performance. Earlier, polyoxometalates (POMs) have shown promise as new potential photocatalysts for degrading organic dyes. However, typical POM salts are water-soluble and can cause secondary pollution if used directly as photocatalysts.

To address this issue, Li et al. developed a novel MOF named

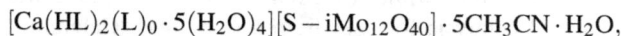

$$[Ca(HL)_2(L)_0 \cdot 5(H_2O)_4][S - iMo_{12}O_{40}] \cdot 5CH_3CN \cdot H_2O,$$

where L = 1,4-bis(pyridinil-4-carboxylato)-1,4-dimethylbenzene, through a slow diffusion reaction of the ligand into $Ca_2[SiMo_{12}O_{40}] \cdot nH_2O$. This MOF retained the photocatalytic properties of POMs while having an insoluble structure, preventing secondary pollution in water. As a heterogeneous photocatalyst, the MOF efficiently catalyzed the degradation of Rhodamine-B (RhB) under UV irradiation. The presence of the MOF significantly enhanced the photocatalytic activity, increasing RhB degradation from 35.84% (without a catalyst) to 91.75% after 90 minutes of UV irradiation shown in Figure 4.7 [43].

4.5.3 DRUG DELIVERY

Ca-MOFs possess remarkable biocompatibility and tunable pore sizes, making them well-suited for drug delivery systems. They have the ability to encapsulate and protect therapeutic agents, enabling controlled and targeted drug release, enhancing drug stability, and reducing side effects. Given their cost-effectiveness, benign nature, abundance in the body, and high recommended consumption, Ca-MOFs hold great promise for various biomedical applications [44]. In a study by Joseph et al., the heterometallic CaSr-BTC MOF demonstrated higher bone mineralization (hydroxyapatite) in pre-osteoblastic cells compared to MOFs containing individual metals

(Ca-BTC and Sr-BTC). This suggests that CaSr-MOFs can release Ca^{2+} and Sr^{2+} ions at an appropriate rate, leading to a cumulative effect on osteoinduction. Further investigations revealed that CaSr-BTC upregulated differentiation markers ALP and BSP II while downregulating COL1, confirming its ability to modulate osteoblast-specific mRNA levels in hMSCs [45].

Additionally, Au et al. introduced DMOG during the synthesis of CaSr-MOFs, creating CaSr-DMOG-MOF to explore increased vascularization potential through VEGF production from hMSCs. Surprisingly, CaSr-DMOG-MOF showed poorer performance than CaSr-BTC in inducing proliferation and differentiation of MC3T3 cells and hMSCs. Another study by Wang et al., focused on a Ca-based nano-MOF called calcium zoledronate (CaZol) coated with polyethylene glycol (PEG) and integrated with folate (Fol)-targeted ligands. This Fol-targeted CaZol nano-MOF acted as a potent anticancer agent, significantly enhancing the direct anti-tumor activity of Zol by 80-85% in vivo. The treatment inhibited tumor neovasculature, cell proliferation, and induced apoptosis [46].

In another study, Li et al., utilized CaZol as a nanocarrier for pDNA (plasmid DNA) delivery. The pH-sensitive CaZol exhibited remarkable stability in the physiological environment (pH 7.4), and it could release the encapsulated pDNA in a weakly acidic environment (pH 5.5). This property ensures better cellular uptake efficiency and desired gene expression efficiency both in vitro and in vivo [47]. Chen et al., developed a pH-responsive nano-MOF, Ca/Pt(IV)@pHisPEG, for drug delivery in nanomedicine applications. These nanoparticles showed extended circulation time and efficient accumulation in tumor tissues at pH 7.4. However, under slightly acidic conditions (pH 6.5), the nano-MOF's imidazole groups underwent protonation, causing the particles to expand in size. This pH-triggered expansion improved tumor retention and cellular uptake of the nanoparticles.

Inside endo/lysosomes, the reduced pH led to the decomposition of the nano-MOF and subsequent drug release, resulting in effective cancer cell elimination. In animal tumor models, the nano-MOF demonstrated impressive efficacy with low drug doses, especially against solid tumors with lower pH levels. This study highlights the potential of pH-responsive nano-MOFs as promising nanomedicine for targeted cancer therapy [48]. Ca-MOFs represent a frontier in drug delivery, offering a multifaceted approach to improving therapeutic outcomes. Their capabilities in drug protection, controlled release, pH responsiveness, and targeted delivery underscore their potential in revolutionizing drug delivery strategies, advancing precision medicine, and minimizing adverse effects in patients.

4.5.4 ENVIRONMENTAL REMEDIATION

The utilization of Ca-MOFs in environmental remediation presents a promising avenue. The use of Ca-MOFs in wastewater treatment holds significant potential. These MOFs can effectively remove dyes, heavy metals, toxins, and pesticides from industrial effluents and municipal wastewater. Additionally, their regenerability allows for sustained use, minimizing waste generation and contributing to a circular economy. Ca-MOFs offer distinctive attributes that make them well-suited for addressing

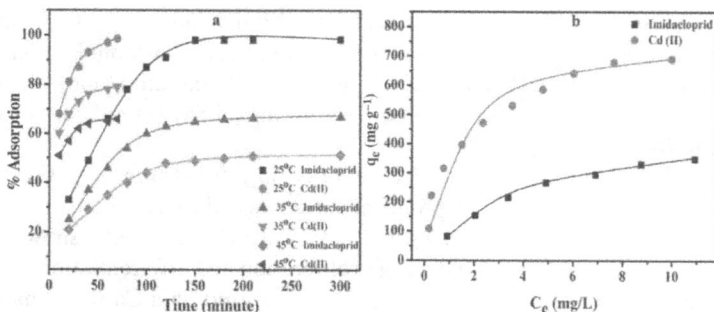

Figure 4.8 (a) The effect of temperature on adsorption of imidacloprid and Cd(II), as temperature increases rate of adsorption decreases, (b) showing the effect of initial concentration of pollutants [50]. (Adapted with permission from Singh *et al. Chemosphere*, 272:129648, 2021. Copyright 2021 Elsevier.)

environmental challenges. Their adjustable structures and substantial porosity enable efficient capturing of pollutants such as heavy metals and organic contaminants from water and air sources [49]. Ca-MOFs serve as effective adsorbents, aiding in the decontamination of polluted environments. Furthermore, the pH-responsive characteristics of Ca-MOFs can be employed for targeted pollutant elimination, as variations in pH trigger the release of trapped contaminants. This trait enhances the efficiency and specificity of remediation procedures. Additionally, by incorporating specific ligands, Ca-MOFs can be tailored to enhance pollutant adsorption effectiveness and selectivity. Altogether, the integration of Ca-MOFs in environmental remediation underscores their potential to contribute to sustainable approaches for pollution mitigation and ecological preservation [36].

The efficient removal of heavy metal ions and pesticides from contaminated water is vital for public health and environmental preservation. A novel single crystal of calcium fumarate $[Ca(C_4H_4O_4)1.5 (H_2O)(CH_3OH)_2]$ was synthesized and characterized using X-ray crystallography, confirming its formation as a 3D MOF structure. These MOFs were then applied for simultaneous removal of imidacloprid, a widely used pesticide, and toxic Cd(II) from water systems. Optimal adsorption occurred at pH 6.5 for imidacloprid and 7.8 for Cd(II). Adsorption data was fitted to various isotherm models, revealing a high adsorption capacity of 467.23 mg g^{-1} for imidacloprid and 781.2 mg g^{-1} for cadmium ions (Figure 4.8). The adsorbent exhibited reusability across multiple cycles. Therefore, the 3D CaFu MOFs present a promising solution for effective removal of pollutants from wastewater [50].

Novel modifications of metal-organic framework (MOF) materials hold considerable promise as efficient adsorbents for phosphate removal from aquatic environments. Wei et al., developed an eco-friendly and highly effective La/Ca composite (La = lanthanum), denoted as La/Ca-BTC, through the calcination of La/Ca MOFs to enhance phosphate adsorption. The incorporation of Ca^{2+} into the composite led to reduced La-release in the presence of humic acid (HA), with a significant reduction of approximately 52.04% compared to La-BTC. Additionally, La/Ca-BTC demonstrated the capability to concurrently eliminate natural organic matter (NOM), offering substantial implications for aquatic remediation strategies. These findings hold significant implications for advancing the field of environmentally friendly phosphate adsorbents, highlighting their potential for sustainable aquatic pollutant mitigation [51]. Ca-MOFs serve as versatile tools in the realm of environmental remediation. Their adsorption, catalytic, and gas capture capabilities position them as potent allies in the quest for cleaner air, water, and soil. By harnessing their distinctive attributes, Ca-MOFs contribute to sustainable solutions that restore and protect the natural environment.

4.6 CONCLUSION

In conclusion, this chapter offers a comprehensive exploration of Ca-MOFs, encompassing their synthesis, characterization, properties, and wide-ranging applications. The attributes and structural chemistry of Ca-MOFs are markedly influenced by the geometry and functional aspects of the organic linkers employed during synthesis, as well as the reaction conditions. The framework's construction involves diverse metal nodes, incorporating primary building units (PBUs) and secondary building units (SBUs) such as dimers, trimers, tetramers, and 1D chains. The adaptability of Ca-MOFs' structural diversity, along with their biocompatibility and eco-friendly features, positions them as promising contenders across various domains. These materials demonstrate remarkable efficacy in gas storage, separation, drug delivery systems, and environmental restoration, among other fields.

Ca-MOFs showcase noteworthy capabilities in separating noble gases and alkane isomers, underlining their potential for industrial applications. Furthermore, their biocompatibility and minimal toxicity render them appealing for drug delivery and biological uses. Incorporating multiple biocompatible metals into Ca-MOFs might optimize drug delivery efficiency and augment their biological functionality. This chapter not only underscores the significant role of Ca-MOFs in addressing contemporary challenges but also elucidates their potential to reshape diverse scientific and industrial realms. It is important to acknowledge that fully harnessing their potential is an ongoing process, necessitating continued research to overcome current limitations and unveil fresh applications.

The ever-evolving field of Ca-MOFs is poised to drive progress in energy, healthcare, and environmental preservation, leveraging their distinctive attributes and versatility. Serving as a valuable resource, this chapter caters to researchers, scientists, and professionals keen on delving into the intricate nuances of Ca-MOFs. By offering this comprehensive analysis, the chapter contributes to a broader comprehension of the prospects, obstacles, and forthcoming directions in the realm of Ca-MOFs.

REFERENCES

1. H. Furukawa, K. E. Cordova, M. OKeeffe, and O. M. Yaghi. The chemistry and applications of metal-organic frameworks. *Science*, **341**(6149):1230444, 2013.

2. H.-C. J. Zhou and S. Kitagawa. Metal–organic frameworks (MOFs). *Chemical Society Reviews*, **43**(16):5415–5418, 2014.

3. K. K. Gangu, S. Maddila, S. B. Mukkamala, and S. B. Jonnalagadda. A review on contemporary metal–organic framework materials. *Inorganica Chimica Acta*, **446**:61–74, 2016.

4. H. Li, L. Li, R.-B. Lin, W. Zhou, Z. Zhang, S. Xiang, and B. Chen. Porous metal-organic frameworks for gas storage and separation: Status and challenges. *EnergyChem*, **1**(1):100006, 2019.

5. T. Tian, Z. Zeng, D. Vulpe, M. E. Casco, G. Divitini, P. A. Midgley, J. Silvestre-Albero, J.-C. Tan, P. Z. Moghadam, and D. Fairen-Jimenez. A sol–gel monolithic metal–organic framework with enhanced methane uptake. *Nature Materials*, **17**(2):174–179, 2018.

6. W. P. Lustig, S. Mukherjee, N. D. Rudd, A. V. Desai, J. Li, and S. K. Ghosh. Metal-organic frameworks: Functional luminescent and photonic materials for sensing applications. *Chemical Society Reviews*, **46**(11):3242–3285, 2017.

7. J. D. Rocca, D. Liu, and W. Lin. Nanoscale metal–organic frameworks for biomedical imaging and drug delivery. *Accounts of Chemical Research*, **44**(10):957–968, 2011.

8. S. Keskin and S. Kızılel. Biomedical applications of metal organic frameworks. *Industrial & Engineering Chemistry Research*, **50**(4):1799–1812, 2011.

9. J. Yang and Y.-W. Yang. Metal–organic frameworks for biomedical applications. *Small*, **16**(10):1906846, 2020.

10. S. Xian, Y. Lin, H. Wang, and J. Li. Calcium-based metal–organic frameworks and their potential applications. *Small*, **17**(22):2005165, 2021.

11. G. Ye, C. Chen, J. Lin, X. Peng, A. Kumar, D. Liu, and J. Liu. Alkali/alkaline earth-based metal–organic frameworks for biomedical applications. *Dalton Transactions*, **50**(47):17438–17454, 2021.

12. P. Horcajada, R. Gref, T. Baati, P. K. Allan, G. Maurin, P. Couvreur, G. Férey, R. E. Morris, and C. Serre. Metal–organic frameworks in biomedicine. *Chemical Reviews*, **112**(2):1232–1268, 2012.

13. J. S. Barbosa, F. Figueira, S. S. Braga, and F. A. Almeida Paz. Metal-organic frameworks for biomedical applications: The case of functional ligands. In *Metal-Organic Frameworks for Biomedical Applications*, 69–92. Elsevier, 2020.

14. X. Chen, Z. Qiao, B. Hou, H. Jiang, W. Gong, J. Dong, H.-Y. Li, Y. Cui, and Y. Liu. Chiral metal-organic frameworks with tunable catalytic selectivity in asymmetric transfer hydrogenation reactions. *Nano Research*, **14**:466–472, 2021.

15. W. Chen and C. Wu. Synthesis, functionalization, and applications of metal–organic frameworks in biomedicine. *Dalton Transactions*, **47**(7):2114–2133, 2018.

16. D. Ozer, N. A. Oztas, D. A. Köse, and O. Şahin. Fabrication and characterization of magnesium and calcium trimesate complexes via ion-exchange and one-pot selfassembly reaction. *Journal of Molecular Structure*, **1156**:353–359, 2018.

17. Y. Sun, Y. Sun, H. Zheng, H. Wang, Y. Han, Y. Yang, and L. Wang. Four calcium(II) coordination polymers based on 2,5-dibromoterephthalic acid and different N-donor organic species: Syntheses, structures, topologies, and luminescence properties. *CrystEngComm*, **18**(44):8664–8671, 2016.

18. E. Gavilan and N. Audebrand. Calcium croconate and calcium oxalato-croconate complexes with 2D and 3D crystal structures. *Polyhedron*, **26**(18):5533–5543, 2007.

19. L. Li, L. Guo, S. Pu, J. Wang, Q. Yang, Z. Zhang, Y. Yang, Q. Ren, S. Alnemrat, and Z. Bao. A calcium-based microporous metal-organic framework for efficient adsorption separation of light hydrocarbons. *Chemical Engineering Journal*, **358**:446–455, 2019.

20. C.-T. Yeh, W.-C. Lin, S.-H. Lo, C.-C. Kao, C.-H. Lin, and C.-C. Yang. Microwave synthesis and gas sorption of calcium and strontium metal–organic frameworks with high thermal stability. *CrystEngComm*, **14**(4):1219–1222, 2012.

21. F. Xiao, W. Gao, H. Wang, Q. Wang, S. Bao, and M. Xu. A new calcium metal organic frameworks (Ca-MOF) for sodium ion batteries. *Materials Letters*, **286**:129264, 2021.

22. P. George, R. K. Das, and P. Chowdhury. Facile microwave synthesis of Ca-BDC metal organic framework for adsorption and controlled release of curcumin. *Microporous and Mesoporous Materials*, **281**:161–171, 2019.

23. D. Kojima, T. Sanada, N. Wada, and K. Kojima. Synthesis, structure, and fluorescence properties of a calcium-based metal–organic framework. *RSC Advances*, **8**(55):31588–31593, 2018.

24. Y. Bian, N. Xiong, and G. Zhu. Technology for the remediation of water pollution: A review on the fabrication of metal-organic frameworks. *Processes*, **6**(8):122, 2018.

25. S. H. Jhung, J. H. Lee, and J. S. Chang. Microwave synthesis of a nanoporous hybrid material, chromium trimesate. Bulletin of the Korean Chemical Society, **26**, 2005.

26. M. Teixeira, R. A. Maia, L. Karmazin, B. Louis, and S. A. Baudron. Ionothermal synthesis of calcium-based metal–organic frameworks in a deep eutectic solvent. *CrystEngComm*, **24**(3):601–608, 2022.

27. T. S. Crickmore, H. B. Sana, H. Mitchell, M. Clark, and D. Bradshaw. Toward sustainable syntheses of Ca-based mofs. *Chemical Communications*, **57**(81):10592–10595, 2021.

28. U. Jamil, A. H. Khoja, R. Liaquat, S. R. Naqvi, W. N. N. W. Omar, and N. A. S. Amin. Copper and calcium-based metal organic framework (MOF) catalyst for biodiesel production from waste cooking oil: A process optimization study. *Energy Conversion and Management*, **215**:112934, 2020.

29. C. B. Jha, C. Singh, R. Varshney, S. Singh, K. Manna, and R. Mathur. Development of novel aspartic acid based calcium bio-MOF designed for management of severe bleeding. *Materials Advances*, **4**, 3330–3343, 2023.

30. B. Tandoğan and N. N. Ulusu. Importance of calcium. *Turkish Journal of Medical Sciences*, **35**(4):197–201, 2005.

31. G. D. Miller, J. K. Jarvis, and L. D. McBean. The importance of meeting calcium needs with foods. *Journal of the American College of Nutrition*, **20**(2):168S–185S, 2001.

32. J. Zhao, M. Wang, H. M. S. Lababidi, H. Al-Adwani, and K. K. Gleason. A review of heterogeneous nucleation of calcium carbonate and control strategies for scale formation in multi-stage flash (MSF) desalination plants. *Desalination*, **442**:75–88, 2018.

33. D. Saha, T. Maity, and S. Koner. Metal–organic frameworks based on alkaline earth metals–hydrothermal synthesis, X-ray structures, gas adsorption, and heterogeneously catalyzed hydrogenation reactions. *European Journal of Inorganic Chemistry*, **2015**(6):1053–1064, 2015.

34. R. K. Alavijeh, K. Akhbari, M. C. Bernini, A. A. G. Blanco, and J. M. White. Design of calcium-based metal–organic frameworks by the solvent effect and computational investigation of their potential as drug carriers. *Crystal Growth & Design*, **22**(5):3154–3162, 2022.

35. Z. Zong, G. Tian, J. Wang, C. Fan, F. Yang, and F. Guo. Recent advances in metal–organic-framework-based nanocarriers for controllable drug delivery and release. *Pharmaceutics*, **14**(12):2790, 2022.

36. F. F. Sukatis, S. Y. Wee, and A. Z. Aris. Potential of biocompatible calcium-based metal-organic frameworks for the removal of endocrine-disrupting compounds in aqueous environments. *Water Research*, **218**:118406, 2022.

37. K. Boukayouht, L. Bazzi, and S. El Hankari. Sustainable synthesis of metalorganic frameworks and their derived materials from organic and inorganic wastes. *Coordination Chemistry Reviews*, **478**:214986, 2023.

38. A. K. Jana, T. Kundu, and S. Natarajan. Stabilization of the anionic metalloligand, $[Ag_6(mna)_6]^{6-}$ (H_2mna = 2-mercapto nicotinic acid), in *cor*, *α-po*, and *sql* nets employing alkaline earth metal ions: Synthesis, structure, and nitroaromatics sensing behavior. *Crystal Growth & Design*, **16**(6):3497–509, 2016.

39. A. Mallick, E.-M. Schön, T. Panda, K. Sreenivas, D. D. Díaz, and R. Banerjee. Fine-tuning the balance between crystallization and gelation and enhancement of CO_2 uptake on functionalized calcium based MOFs and metallogels. *Journal of Materials Chemistry*, **22**(30):14951–14963, 2012.

40. R.-B. Lin, L. Li, H.-L. Zhou, H. Wu, C. He, S. Li, R. Krishna, J. Li, W. Zhou, and B. Chen. Molecular sieving of ethylene from ethane using a rigid metal–organic framework. *Nature Materials*, **17**(12):1128–1133, 2018.

41. A. E. P. Prats, V. A. de la Peña-O'Shea, M. Iglesias, N. Snejko, A. Monge, and E. Gutiérrez-Puebla. Heterogeneous catalysis with alkaline-earth metal-based mofs: A green calcium catalyst. *ChemCatChem*, **2**(2):147–149, 2010.

42. M. Albert-Soriano, P. Trillo, T. Soler, and I. M. Pastor. Versatile barium and calcium imidazolium-dicarboxylate heterogeneous catalysts in quinoline synthesis. *European Journal of Organic Chemistry*, **2017**(43):6375–6381, 2017.

43. X.-L. Hao, Y.-Y. Ma, Y.-H. Wang, W.-Z. Zhou, and Y.-G. Li. New organic–inorganic hybrid assemblies based on metal–bis(betaine) coordination complexes and keggin-type polyoxometalates. *Inorganic Chemistry Communications*, **41**:19–24, 2014.

44. R. P. Heaney. Vitamin d and calcium interactions: Functional outcomes. *The American Journal of Clinical Nutrition*, **88**(2):541S–544S, 2008.

45. N. Joseph, H. D. Lawson, K. J. Overholt, K. Damodaran, R. Gottardi, A. P. Acharya, and S. R. Little. Synthesis and characterization of casr-metal-organic frameworks for biodegradable orthopedic applications. *Scientific Reports*, **9**(1):13024, 2019.

46. K. M. Au, A. Satterlee, Y. Min, X. Tian, Y. S. Kim, J. M. Caster, L. Zhang, T. Zhang, L. Huang, and A. Z. Wang. Folate-targeted ph-responsive calcium zoledronate nanoscale metal-organic frameworks: Turning a bone antiresorptive agent into an anticancer therapeutic. *Biomaterials*, **82**:178–193, 2016.

47. W. Li, X. Xin, S. Jing, X. Zhang, K. Chen, D. Chen, and H. Hu. Organic metal complexes based on zoledronate–calcium: A potential pdna delivery system. *Journal of Materials Chemistry B*, **5**(8):1601–1610, 2017.

48. Y. Yang, L. Xu, W. Zhu, L. Feng, J. Liu, Q. Chen, Z. Dong, J. Zhao, Z. Liu, and M. Chen. One-pot synthesis of ph-responsive chargeswitchable pegylated nanoscale coordination polymers for improved cancer therapy. *Biomaterials*, **156**:121–133, 2018.

49. H. Musarurwa and N. T. Tavengwa. Smart metal-organic framework (MOF) composites and their applications in environmental remediation. *Materials Today Communications*, **33**:104823, 2022.

50. S. Singh, S. Kaushal, J. Kaur, G. Kaur, S. K. Mittal, and P. P. Singh. CaFu MOF as an efficient adsorbent for simultaneous removal of imidacloprid pesticide and cadmium ions from wastewater. *Chemosphere*, **272**:129648, 2021.

51. L. Wei, M. Li, Y. Zhang, and Q. Zhang. The role of Ca^{2+} in the improvement of phosphate adsorption in natural waters: Establishing an environmentally friendly La/Ca bimetallic organic framework. *Environmental Research*, **219**:115126, 2023.

5 Calcium-Based Hydroxide
Synthesis and Applications

Ritika Charak, Anjali Bhardwaj, and
Sanjeev Gautam
Advanced Functional Materials Laboratory, Dr. S.S. Bhatnagar
University Institute of Chemical Engineering and Technology
Panjab University, Chandigarh, India

5.1 INTRODUCTION

Calcium hydroxide, which is commonly known as hydrated lime or slaked lime, is a highly versatile chemical compound with chemical formula $Ca(OH)_2$. The chemical is a white and alkaline solid which is available in powder form as well as dispersion (suspension) form. It is highly soluble in glycerol, slightly soluble in water and completely insoluble in alcohol [1]. Slaked lime has hexagonal structure and CAS number of 1305-62-0. Most common physical and chemical properties of calcium hydroxide are listed in Table 5.1.

These properties are determined by the help of various analysis techniques. Techniques (Figure 5.1) such as X-ray diffraction (XRD) and Transmission electron microscopy (TEM) are used to identify the composition and size of the grain or particle in a compound. These values give the basic information on the physical and chemical properties of a compound. Further, Fourier-transform infrared spectroscopy (FTIR), is used to study the detailed composition of any material, as it identifies of unknown materials and confirmation of production materials (incoming or outgoing calcium carbonate($CaCO_3$) is used as a raw material to produce calcium hydroxide which is extracted from minerals such as, limestone and dolomite. The extracted calcium carbonate is then treated with high heat and calcined into calcium oxide(CaO). The most common technique to synthesize calcium hydroxide is by an exothermic reaction of calcium oxide(CaO) with water (H_2O) which is known as slaking of lime. Calcium hydroxide in recent years have also been produced through a range of techniques including chemical precipitation [4], sonochemical reactions [5], sol-gel processes [6], hydrogen plasma-metal reactions [7], heterogeneous phase synthesis [8], and solvothermal reactions [9] etc. By adjusting the synthesis methods and manipulating experimental parameters, it is possible to alter the characteristics of calcium hydroxide for different applications. When the produced calcium hydroxide is mixed

DOI: 10.1201/9781003360599-5

Table 5.1
Physical and Chemical Properties of Calcium Hydroxide

Shape	Hexagonal
Molecular formula	$Ca(OH)_2$
CAS number	1305-62-0 nm
Color	white, light grey
Smell	Oderless
Molecular weight	74.09 g/mol
Density	2.21 mg/m^3
pH value	12–13
Melting point	580 °C
Boiling point	Decomposes before reaching a boiling point
Base	Highly basic
Reactivity	Highly reactive carbon dioxide(CO_2)
Solubility	Soluble in glycerol, water, mineral oils, etc. Insoluble in ethanol
Chemical safety	Corrosive and an irritant

Source: References [1, 2].

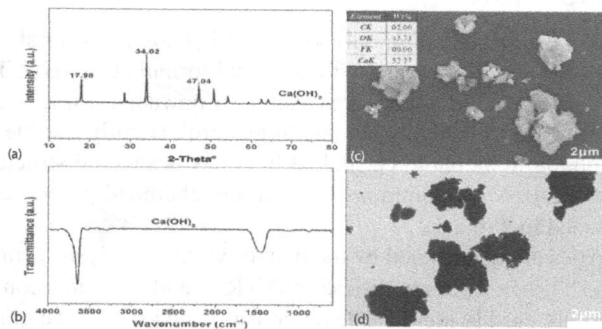

Figure 5.1 Characterization of calcium hydroxide by (a) XRD. (b) SEM. (c) FT-IR, and (d) TEM [3]. (Available via license: Creative Commons Attribution-NonCommercial 3.0 Unported.)

with water, the suspended solution formed is called the milk of lime. When the milk of lime settles down, the clear solution is known as lime water and the settled material is known as lime putty. The lime putty is separated from the lime water and is allowed to age for a while, which is used masonry and lime based mortars.

Slaked lime has been used as a building and construction material since ancient roman civilization [10]. It is also used in treating water and pH regulating agent. Calcium hydroxide is as to synthesize other calcium based materials such as calcium oxide, calcium phosphate, calcium chloride etc. The chemical is used in the food and agriculture sector as a non toxic fertilizer and also as a pickling agent. It is also used to make corn tortillas and as an antibacterial agent [11, 12]. Calcium hydroxide is used in the medical industry as an astringent to kill surface micro-organisms from abscesses and stimulate mineralisation. It is also used as an antacid and as a important part in endodontics [13]. Such vast variety of application makes the chemical to produced in massive range throughout the world where Germany is the biggest exporter of calcium hydroxide with 103 million kilograms being exported and UK is the second with 83 million kilograms of slaked lime being exported in 2018 [14].

Bulk hydrated lime is stored within enclosed silos to mitigate its interaction with atmospheric carbon dioxide and to manage the emission of dust particles. For transportation, specialized air pressure discharge vehicles are employed for both road and rail transit [15]. While a notable portion of hydrated lime is marketed in 25 kg paper sacks, the usage of intermediate bulk containers, capable of holding up to 1 ton, is progressively growing. A considerable global market exists for packaged hydrated lime. The flow characteristics of hydrated lime are notably variable, indicated by an angle of repose that can fluctuate between 0 and 808. This variability hinges on factors such as the presence of entrained air, excess water content, and other influencing elements. Silos may necessitate the incorporation of apparatuses to manage instances of powder bridging above discharge points and to stimulate consistent material flow. These apparatuses can include bin activators, air pads, and, in extreme scenarios, air cannons [16]. Hydrated lime packaged in paper sacks is typically situated on pallets and stored beneath protective cover. By integrating impermeable slip-sheets beneath the bags and securely shrink-wrapping a cover over the loaded pallet, hydrated lime can be stored outdoors for extended durations, potentially spanning several months [17].

5.2 SYNTHESIS METHODS

5.2.1 SLAKING OF LIME

The process of adding water to calcium oxide (CaO) to produce calcium hydroxide is commonly known as lime slaking or the hydration process. When calcium oxide, often referred to as quicklime, reacts with water, it releases heat due to its exothermic nature. The degree of hydration or the ratio of CaO to water has notable effects on the resulting product.

When the hydration process is carried out with precise water quantities, it is termed as "Dry Hydration". In this scenario, the resulting hydrate takes the form of a dry powder. Conversely, when an excess of water is used for hydration, the process is termed as "Slaking". This leads to the creation of a slurry like mixture of calcium hydroxide suspended in water. Notably, lime manufacturers often opt for the dry hydration process to produce powdered hydrated lime, which is widely used across various industries. The chemical formula for this process is as follows:

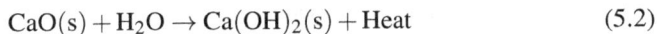

$$CaCO_3 + Heat \rightarrow CaO + CO_2 \qquad (5.1)$$

$$CaO(s) + H_2O \rightarrow Ca(OH)_2(s) + Heat \qquad (5.2)$$

Produced calcium hydroxide can also turn back to calcium carbonate by introduction of atmospheric carbon dioxide this process is called as lime cycle (Figure 5.2). For the purpose of this discussion, our focus is specifically on the lime slaking

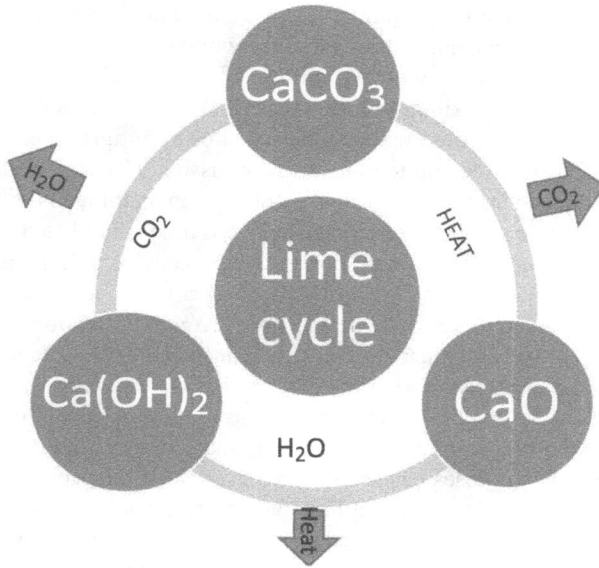

Figure 5.2 Representation of a lime cycle.

process. In slaking, a substantial surplus of water is typically utilized, with ratios ranging from 2.5 parts water to one part CaO to as much as six parts water to one part CaO. This excess water facilitates control over the temperature rise during the exothermic reaction, ensuring a safer and controlled process [18].

While the dry hydration method is suitable for producing powdered hydrated lime, the slaking process finds its application in producing a slurry that serves various industrial purposes, including pollution control and construction. This distinction in methods emphasizes the significance of water quantities in shaping the properties and applications of the resulting calcium hydroxide products. Harper et al. [19] in their article released in 1999, emphasizes the importance of temperature regulation and quantity of water added in the process of slaking. The discussion has been beneficial in preventing heat related damage to the process and enlisting different uses of slaked lime by changing the hydration quantity.

5.2.2 SOL-GEL AND SOLVOTHERMAL PROCESS

The sol-gel/solvothermal method process (Figure 5.3) involves initial synthesizing of precursor calcium alkoxides (methoxides, ethoxides, propoxides, or isopropoxides) through a reaction, where the metallic Ca reacts with the corresponding alcohol at temperatures exceeding 60 °C [20]:

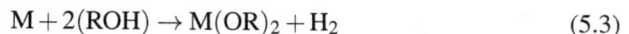

$$M + 2(ROH) \rightarrow M(OR)_2 + H_2 \tag{5.3}$$

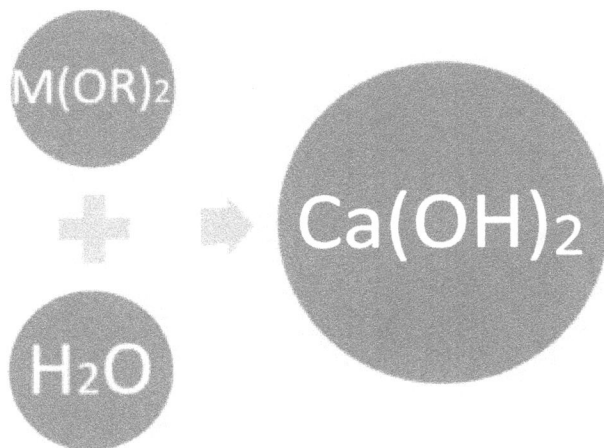

Figure 5.3 Representation of a sol-gel/solvothermal process.

In this context, M represents an alkaline-earth metal, often Ca, Mg, Sr, or Ba, while R typically denotes a methyl, ethyl, or n-propyl group. However, it is worth noting that the release of hydrogen gas during reaction presents a potential hazard. This formation occurs through the hydrolysis reaction:

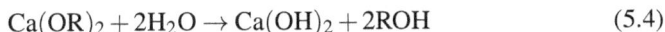

$$Ca(OR)_2 + 2H_2O \rightarrow Ca(OH)_2 + 2ROH \qquad (5.4)$$

This hydrolysis reaction could be initiated before the treatment application, leading to the creation of nanoscale $Ca(OH)_2$ particles. Subsequently, these particles could be applied as a dispersion in alcohol. Poggi et al. [9] uses similar technique process for the synthesis of calcium hydroxide. The first step involved mixing 10 g of granular calcium to 500 mL of alcohol were placed inside a high-pressure reactor. In order to hydrolyze the prepared precursor to calcium hydroxide, 35 mL of water is added inside a reaction chamber in a nitrogen atmosphere at 70 °C for 60 minute. Camerini et al. [21] also used a similar route of synthesis of calcium hydroxide.

5.2.3 CHEMICAL PRECIPITATION IN AQUEOUS MEDIUM

Chemical precipitation is a simple operation, cost effective method for removal of metallic ions or cations to an insoluble form (Figure 5.4). Through this process, most of the metals precipitated are hydroxides. Mixing calcium chloride with sodium hydroxide eventually leads cationic removal of sodium atom, where calcium hydroxide and sodium chloride are produced as the end products. Darroudi et al. [6] used this approach of synthesis by reacting calcium chloride ($CaCl_2$) with sodium hydroxide(NaOH) using sol-gel method.

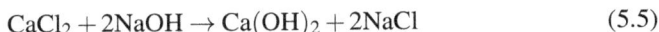

$$CaCl_2 + 2NaOH \rightarrow Ca(OH)_2 + 2NaCl \qquad (5.5)$$

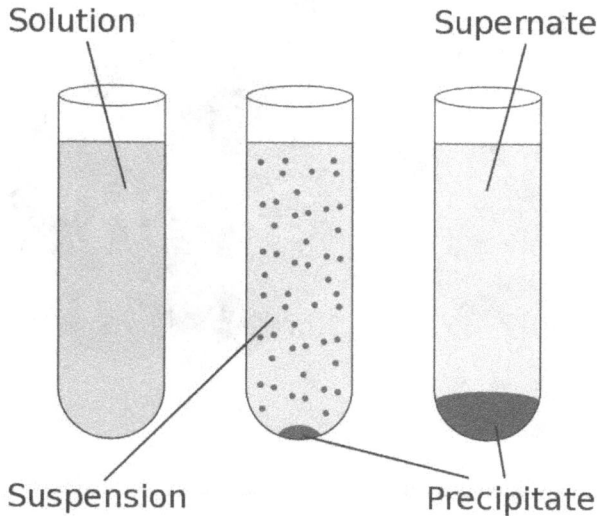

Figure 5.4 Representation of chemical precipitation. CC0 1.0 Universal (CC0 1.0). (Public Domain Dedication.)

On the other hand Aniruddha et al. [4] used calcium nitrate dihydrate [Ca(NO$_3$)$_2$.2H$_2$O] and sodium hydroxide (NaOH) for the synthesis of calcium hydroxide through chemical precipitation method. The article argues that calcium nitrate dihydrate is a better precursor than calcium chloride due to higher solubility in water which leads to quick formation of nano crystals and the process is less energy intensive due to the precursor having low precipitation temperature.

5.2.4 THE HYDROGEN PLASMA-METAL REACTION

The Hydrogen Plasma-Metal Reaction (HP-MR) method is an innovative vapor deposition technique that offers a cost-effective means for industrially producing nanoparticles characterized by high purity and excellent crystalline quality. Unlike conventional physical deposition methods, such as vacuum evaporation, which rely on thermal effects and are limited to elements with low melting points and significant saturation vapor pressures.

In the HP-MR method, an arc plasma generates remarkably high temperatures of up to 10,000 °C. Notably, this process induces the decomposition of hydrogen molecules into atoms or ions, which in turn exhibit heightened solubility within liquid metals compared to their normal gaseous state. This dissolved hydrogen subsequently recombines into gas form. Consequently, the resulting supersaturated gas swiftly escapes from the liquid metal, leading to a pronounced acceleration of the metal's evaporation process. The unique interplay of these factors forms the foundation of the HP-MR approach.

Remarkably, this methodology has facilitated the production of a diverse array of alloys and intermetallic nanoparticles, all meticulously crafted using the HP-MR technique. Notably, the size of these nanoparticles can be precisely adjusted by manipulating variables such as hydrogen content and current strength [7]. Recent advancements have extended the capabilities of the HP-MR method, revealing its efficacy in synthesizing oxide nanoparticles derived from rare earth metals. Liu et al. [7] successfully synthesizes calcium hydroxide using HP-MR technique, which gives high yield and purity nanoparticles.

5.3 APPLICATIONS

5.3.1 IRON AND STEEL INDUSTRY

In the context of the Bayer process for alumina production, slaked lime plays a pivotal role in regenerating sodium hydroxide from sodium carbonate solutions (as mentioned earlier). Additionally, it finds application in flotation procedures aimed at enhancing copper ore, as well as in the extraction of gold and silver. It holds vital significance in the extraction of uranium from gold slimes and in the recovery of nickel and tungsten post-smelting processes [17].

Moreover, slaked lime finds utility in the production of magnesia and magnesium metal. Within seawater processes, a high-calcium hydrate is utilized to precipitate magnesium hydroxide. Notably, the Dow natural brine process employs dolomitic milk of lime for this purpose. Kariya et al. [22] uses a new composite material for chemical heat storage (CHS) systems, using calcium oxide/water/calcium hydroxide $(CaO/H_2O/Ca(OH)_2)$ reaction. In this context, the introduction of blends involving expanded graphite (EG) and calcium hydroxide has been discovered to enhance both reaction efficiency and the material's formability. These attributes are of significant importance for the utilization of such systems within packed-bed CHS heat exchangers. In terms of performance, a blend comprising 11 wt% EG showcased notable results. Its maximum mean heat output reached 1.76 kW (per kilogram of material), a figure twice as substantial as that achieved by pure $Ca(OH)_2$ (0.85 kW per kg of material).

5.3.2 MEDICAL FIELD

Calcium hydroxide has been in use for more than a century in medical field as a pulp capping material. The chemical is used mostly in the filed of dentistry due to its anti-microbal and disinfectant properties. Calcium hydroxide has been in use as a pH adjuster and acid regulator in various industries, so the chemical can also be useful as a mild antacid, used for ingestion or heatburn inside the stomach.

5.3.2.1 In Endodontics

In the realm of endodontics, calcium hydroxide has gained significant popularity owing to its numerous benefits. Calcium hydroxide was first introduced over a century ago by Herman to dentistry as a pulp-capping material but today it is used widely in the field of endodontics [23]. One of its primary advantages lies in its initial bactericidal impact, which subsequently transitions into a bacteriostatic effect. This dual action contributes to improved bacterial control. Additionally, calcium hydroxide plays a pivotal role in fostering healing and repair processes within the treated area. Its high pH level effectively stimulates the activity of fibroblasts, further aiding in tissue restoration (Figure 5.5).

Another noteworthy benefit is its capability to halt internal resorption, preventing the degradation of tooth structure. Furthermore, calcium hydroxide acts as a neutralizing agent, counteracting the deleterious effects of low pH acids that may be present. It also boasts affordability and user-friendliness, making it a practical choice for endodontic applications. Zahed Mohammadi et al. [24] in their study conclude that the combinations of calcium hydroxide with camphorated paramonochlorophenol or CHX can be potentially used in effective intracanal medicaments on cases with fungal infections.

However, alongside its advantages, calcium hydroxide does exhibit certain drawbacks. It does not exclusively induce dentinogenesis, the process of new dentin formation. Instead, its influence primarily leads to the development of reparative dentin, which might not possess the same structural integrity as the original dentin. Moreover, there is an association with resorption of primary teeth, which can be a concern in certain cases.

Over time, calcium hydroxide can exhibit limitations, including potential dissolution at the cavosurface after a year. It is susceptible to degradation under tooth flexure, which could affect its long-term stability. Additionally, marginal failures may occur when subjected to amalgam condensation, possibly compromising the treatment's effectiveness. It's important to note that calcium hydroxide lacks strong adhesive properties to dentin or resin restorations, potentially impacting its longevity and reliability when used in conjunction with these materials [25]. Studies done by

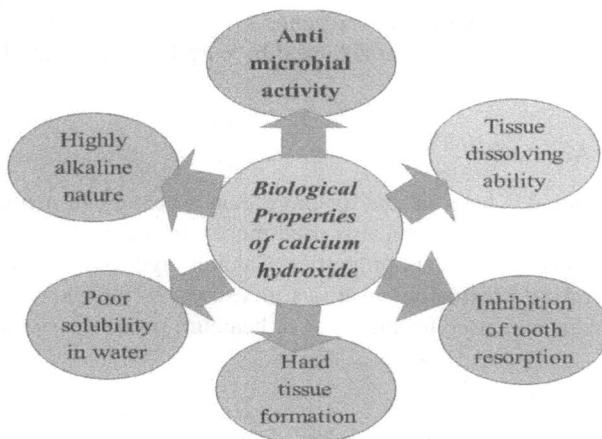

Figure 5.5   Application of calcium hydroxide in endodontics.

Sahebi et al. [26] shows that short term root canal filling with Ca(OH)$_2$ reduced the strength of the dentin of mature human teeth by 15 percent in a month. Dr. Gluskin et al. [27] in their article concludes on taking precautions towards overfilling of medication or obturation materials in endodontics can cause permanent neural damage.

5.3.2.2 As an Antacid

An antacid is a medicine used to neutralize the acid present in the stomach which prevents ingestion or heatburn inside the stomach. Antacids can quickly relieve the symptoms but are not capable of treating underlying cause of a problem. Calcium hydroxide can be used as an antacid to reduce acidity of hydrochloric acid (HCl) present inside the stomach lining. A simple reaction with the acid gives calcium chloride and water, which are chemically stable compounds. Erikson et al. [28] compares the antacid effect of various calcium, aluminium and magnesium based salts on the pH vs time graph with performance criterias such as delay time which is the time needed for the antacid to raise the pH of a 0.1 N HCl solution to an equilibrium value, static equilibrium (pH) is the equilibrium value beyond which no further HCl can be added to the system and dynamic equilibrium (pH), which is the approximate equilibrium value resulting (in excess antacid conditions) when acid was added c at a rate equal to 4 ml of 0.1 N HCl per minute. The studies showed that calcium hydroxide is a mildly effective pH regulator and can be used as an antacid.

5.3.3 BUILDING AND CONSTRUCTION INDUSTRY

5.3.3.1 Building Materials

1. **Asphalt pavement:** Lime is used in stabilization of soil for construction of roads. Lime is the best choice over all other additive both for long term cost saving, and for improving shelf life durability of roads because it is a superior anti-stripping additive.
2. **Sand-lime bricks:** Sand-lime bricks have a notable market presence in Europe. These bricks are crafted by blending slaked lime with sand and subsequently subjecting the mixture to autoclaving for multiple hours under steam pressure at approximately 180°C and 1 MPa. In certain procedures, it's crucial for the lime to exhibit non-expansive properties. This is imperative to prevent expansion within the autoclave, which could lead to the formation of excessively large bricks.
3. **Mortars:** In the realm of mortars, slaked lime finds utility in lime-cement-sand compositions of varying ratios. By adjusting the mixture's composition, the desired compressive strength of the mortar is achieved (34). The inclusion of lime offers numerous advantages. It enhances the workability and moisture retention of the fresh mortar, while also bolstering the adhesive strength between the mortar and the masonry. As a result, the mortar gains improved flexibility and becomes less susceptible to cracking.

5.3.3.2 Deacidification of Paper and Canvas

Through time the cellulose present in the paper starts to age. Due to promoted acid pH in old artifacts, the cellulose fibers start to depolymerize. This can solved by paper deacidification which is a important technique in conservation and restoration of cultural heritage. Calcium hydroxide posses the best deacidification properties, which when applied deacidify the paper and reacts with the carbon dioxide present in the air and makes a calcium carbonate layer over the applied region. It has helped to restore damaged artifacts as well as remove yellowish acid residue coming from the wood pulp and rag from the paintings (Figure 5.6). Giorgi et al. [29] finds that calcium hydroxide is physio-chemically compatible with almost all kind of papers and the calcium carbonate film deposition on the paintings make them last longer. Emiliano Carretti et al. [31] studies the interactions between acrylic copolymers and calcium hydroxide leading them to conclude that the performance of calcium hydroxide is not affected by the presence of copolymer and can be used as a part of restoration if used with calcium hydroxide to enhance physio-chemical compatibility with the paper.

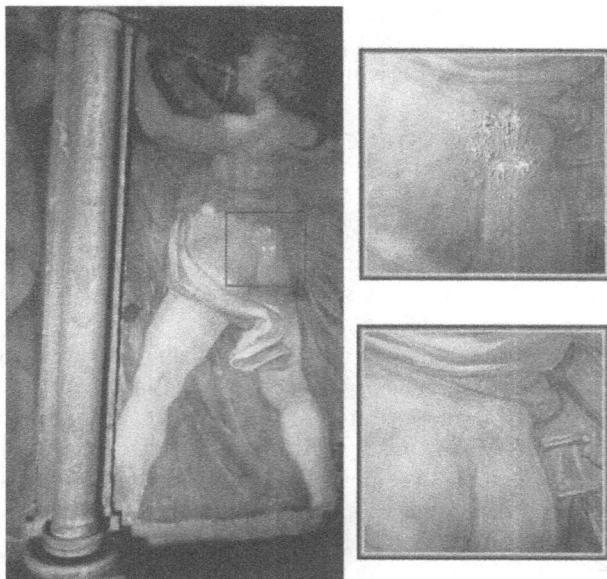

Figure 5.6 Utilizing $Ca(OH)_2$ submicrometer crystals dispersed in propan-1-ol, conducted a preconsolidation treatment in the restoration process of Santi di Tito's 16th-century wall painting titled "Gli Angeli Musicanti" present in Counterfaçade of the Santa Maria del Fiore Cathedral in Florence. The boxed area, highlighted in the images, underwent a significant change: the top image captures its original state prior to restoration, while the bottom image exhibits appearance after restoration procedure. (Adapted with permission from Ambrosi *et al.* [30]. Copyright 2001, American Chemical Society.)

This insight is valuable for guiding conservation efforts and ensuring the long-term preservation of cultural artifacts.

5.3.4 WATER TREATMENT

5.3.4.1 Purification of Water

Slaked lime finds utility in the elimination of both temporary and permanent water hardness. When calcium hydroxide reacts with calcium and magnesium hydrogen carbonates, it yields insoluble calcium carbonate and magnesium hydroxide as products:

$$Ca(HCO_3)_2 + Ca(OH)_2 \rightarrow 2CaCO_3 + 2H_2O \qquad (5.6)$$

$$Mg(HCO_3)_2 + 2Ca(OH)_2 \rightarrow Mg(OH)_2 + 2CaCO_3 + 2H_2O \qquad (5.7)$$

For the elimination of permanent hardness, which is attributed to calcium and magnesium chlorides and sulfates, a combination of hydrated lime and sodium carbonate is employed:

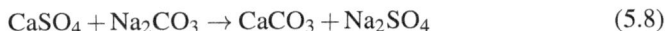

$$CaSO_4 + Na_2CO_3 \rightarrow CaCO_3 + Na_2SO_4 \qquad (5.8)$$

$$MgSO_4 + Ca(OH)_2 + Na_2CO_3 \rightarrow Mg(OH)_2 + CaCO_3 + Na_2SO_4 \qquad (5.9)$$

Purification of water involves raising the pH level to a value greater than 11 for a duration of 1–2 days. This is followed by recarbonation to restore the pH to the range of 8–9. This treatment not only effectively eradicates bacteria but also eliminates temporary water hardness. Conversely, acidic water can be neutralized through the addition of lime [32].

5.3.4.2 Softening and remineralisation of water

Hydrated lime is used to remove carbonate hardness from the water. This hardness is caused by magnesium and calcium salts; for the most part it is treated by means of the lime-soda process. This entails the precipitation of magnesium by lime. The calcium salt that is co-produced reacts with the soda ash to form a calcium carbonate precipitate. Lime enhanced softening can also be used to remove arsenic from water [33]. Hydrated lime is also used in the Nalgonda technique of defluoridation and is based on combined use of alum and lime in a two-step process and has been claimed as the most effective technique for fluoride removal [34].

5.3.4.3 Mitigation of Acidic Effluents

The prevailing approach for addressing acidic effluents involves an active treatment process employing a chemical neutralizing agent. In this method, an alkaline substance is introduced to counteract the acidity. This addition of an alkaline agent to sites affected by Acid Mine Drainage (AMD) results in a pH elevation, which, in turn, expedites the chemical oxidation of ferrous iron. This alkaline intervention also triggers the precipitation of many metals in the solution as hydroxides and carbonates. Consequently, an iron-rich sludge is generated, potentially containing diverse other metals depending on the composition of the treated mine water. Several neutralizing agents have been employed for this purpose. Lime (calcium oxide), slaked lime, calcium carbonate, sodium hydroxide, sodium carbonate, magnesium hydroxide, and magnesium oxide are among the options [32]. These agents vary in their efficacy and cost. For instance, sodium hydroxide exhibits higher effectiveness but comes at a cost approximately nine times greater than lime. Moreover, the use of calcium-containing neutralizing reagents facilitates the removal of sulfate as gypsum. While chemical neutralization proves effective in remediating AMD, it introduces a challenge associated with the disposal of the resulting bulky sludge [35].

5.3.5 FLUE GAS DESULFURIZATION

Within the realm of flue gas desulfurization (FGD), several distinct processes make use of slaked lime (calcium hydroxide, $Ca(OH)_2$), as outlined below:

1. **Dry Scrubbing:** There are two main dry scrubbing processes:-
   i. *Dry injection system:-* In this system dry hydrated lime is directly injected in Flue gas duct to remove SO_2 and HCl.
   ii. *Spray dryer system:-* In this system, lime is injected dry hydrated lime in an atomized slurry form in to another vessel. The shape of spray dryer is like "silo" (cylindrical top and conical bottom) from the top slurry is sprayed through an atomizer (e.g, nozzles) in order to absorbs sulphur di oxide and HCl. Water present in lime slurry is then evaporated by the hot gas. The scrubbed Flue gas is then followed to a horizontal duct and the excess of the dry, unreacted lime and its product is sent to the bottom of the cone from where it is removed. Both dry injection system and spray dryer system yields final product in dry phase which is sent to the particulate controlling device for the further processing. Recently, dry scrubbing method improvised become more efficient in removing impurities.
2. **Wet Scrubbing:** In this system, lime is firstly treated with water and form slurry which is sprayed into Flue gas scrubber. In typical system the gas which are to be cleaned is allowed to enter through the bottom in a cylindrical tower and treated with lime shower while going up. And sulphur di oxide absorbs the spray and converted into Calcium sulphite(ppt).
   The sulphite can be converted into gypsum (a saleable by-products). Hence highly alkaline lime is required to remove sulphur di oxide with greater efficiency and reduce scaling potential. So magnesium enhance lime is generally preferred for wet scrubbing [36].

5.3.6 AGRICULTURAL INDUSTRY

5.3.6.1 Soil pH Stabilizer

The benefits of soil pH adjustment through the use of limestone have been outlined in the "Uses and Specifications of Limestone" section. However, hydrated lime offers an added advantage due to its rapid reactivity. Pettry et al. [37] studied the addition of calcium hydroxide and its use in amendment of swelling and shrinking in soil.

1. Bleaching processes: In the realm of bleaching, a chemical reaction involving hydrated lime and chlorine results in the creation of a powdered mixture called bleaching powder. This mixture contains calcium hypochlorite and calcium chloride.

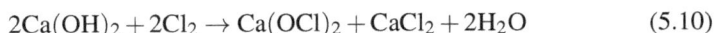

$$2Ca(OH)_2 + 2Cl_2 \rightarrow Ca(OCl)_2 + CaCl_2 + 2H_2O \qquad (5.10)$$

Calcium hypochlorite solution, derived from this reaction, plays a pivotal role in bleaching wood pulp [38].

2. Production of precipitated calcium carbonate (PCC): Precipitated calcium carbonate (PCC) is generated by directing gases containing carbon dioxide through milk of lime. By carefully controlling the conditions, a finely divided calcium carbonate with exceptional reflectivity and a median particle size of 0.02–0.2 μm is synthesized. Some variants of PCC are coated with compounds that enhance their compatibility with organic substances, such as plastics and rubber [17].

5.3.6.2 Canning and Preserving Food

Food grade calcium hydroxide or pickling lime is used for canning and preservation of food. Pickling lime or food grade calcium hydroxide is safe for consumption while industry level slaked lime when consumed will cause calcium hydroxide poisoning. Some of the symptoms of calcium hydroxide poisoning include: vision loss, severe pain or swelling in your throat, burning sensation on your lips or tongue, burning sensation in your nose, eyes, or ears, difficulty breathing, abdominal pain, nausea, and vomiting, blood in the stool, loss of consciousness, low blood pressure, low blood acidity, etc. Introduction of pickling lime in the canning or pickling process makes the vegetables crunchy, as the calcium in pickling lime binds with the pectin, which makes the vegetables firmer [39] as shown in Figure 5.7. Due to its alkaline nature it may neutralize the acid, which is used in canning process to kill bacteria. Thus,

Figure 5.7 Pickling of food by the help of calcium hydroxide.

the lime used should be washed thoroughly before the canning process. Calcium hydroxide can be used as a preservative in ways such as.

Corn products: Native American use Calcium hydroxide in nixtamalization of corn and other grains. Well in Nahuatl the language of Aztec the word calcium hydroxide is pronounced as nextli, when maize is treated with nextli to become nixtamal(hominy) and the process called nixtamalization, which is used increase availability of niacin (vitamin B3), enhance taste, makes the digestion of corn easier not only it enhance nutritional value but it also reduce mycotoxin present in grains up to 97%–100% [40]. In modern times, the majority of items made using corn flour (masa harina), including tortillas, sopes, and tamales, incorporate calcium hydroxide as an ingredient.

Sugar refining: Certain types of sugar use calcium hydroxide in its production process. During the process of carbonation, calcium hydroxide is incorporated in the sugar to improve the stability and remove impurities from the sugar. The sugar that use calcium hydroxide are as follows.

1. **Sugar Beet Refining:** The refining of sugar beet involves treating the crude sugar solution with milk of lime. This step precipitates calcium salts of organic and phosphoric acids. Subsequent neutralization with carbon dioxide leads to the precipitation of calcium carbonate. After filtration to remove solids, a purified sugar solution is obtained. Given the substantial quantities of lime (approximately 250 kg/t of sugar) and carbon dioxide required, sugar beet producers often integrate lime kilns into their processes.
2. **Sugar Cane Refining:** The refining process for sugar cane requires notably less hydrated lime (typically around 5 kg/t) compared to sugar beet refining. This distinction arises from the inherent purity of raw sugar extracted from sugar cane as compared to sugar beet [17].

Fruit juices: Some fortified juices add calcium to their product to boosts its nutritional value without compromising the taste or appearance of the product. Addition of calcium hydroxide is one such way to add the needed calcium. This addition is strictly regulated as calcium can only be used as a bait for making products look healthy [41]. These regulations can be different from country to country or may depend upon the consumption of a particular product and the consumer's age.

5.4 CONCLUSION

Calcium hydroxide has already demonstrated its application in numerous field. We may foresee future evolution in upgrading the use of nanolime, and development in the area continues. These development will help to makes the life easy for the mankind. Emerging uses of nano technology, remarkable potential of calcium hydroxide and growing interest of researchers in this field is forcing us to believe that the development in the uses of calcium hydroxide is going to be proceed very soon. We have already seen a lot of uses of calcium hydroxide or nano lime and studies says there are a lot to discover yet.

REFERENCES

1. United States National Institutes of Health. Compound summary. Slaked lime. 2023. Retrieved from https://pubchem.ncbi.nlm.nih.gov/compound/Slaked-lime.
2. Chemicals learning. Calcium hydroxide: Definition, formula, molar mass, uses and properties, 2021.
3. Z. Liang, Q. Wang, B. Dong, B. Jiang, and F. Xing. Ion-triggered calcium hydroxide microcapsules for enhanced corrosion resistance of steel bars. *RSC Advances*, **8**(69):39536–39544, 2018.
4. A. Samanta, D. K. Chanda, P. S. Das, J. Ghosh, A. K. Mukhopadhyay, and A. Dey. Synthesis of nano calcium hydroxide in aqueous medium. *Journal of the American Ceramic Society*, **99**(3):787–795, 2016.
5. M. A. Alavi and A. Morsali. Ultrasonic-assisted synthesis of $Ca(OH)_2$ and CaO nanostructures. *Journal of Experimental Nanoscience*, **5**(2):93–105, 2010.
6. M. Darroudi, M. Bagherpour, H. A. Hosseini, and M. Ebrahimi. Biopolymer-assisted green synthesis and characterization of calcium hydroxide nanoparticles. *Ceramics International*, **42**(3):3816–3819, 2016.
7. T. Liu, Y. Zhu, X. Zhang, T. Zhang, T. Zhang, and X. Li. Synthesis and characterization of calcium hydroxide nanoparticles by hydrogen plasma-metal reaction method. *Materials Letters*, **64**(23):2575–2577, 2010.
8. C. Rodriguez-Navarro, A. Suzuki, and E. Ruiz-Agudo. Alcohol dispersions of calcium hydroxide nanoparticles for stone conservation. *Langmuir*, **29**(36):11457–11470, 2013.
9. G. Poggi, N. Toccafondi, D. Chelazzi, P. Canton, R. Giorgi, and P. Baglioni. Calcium hydroxide nanoparticles from solvothermal reaction for the deacidification of degraded waterlogged wood. *Journal of Colloid and Interface Science*, **473**:1–8, 2016.
10. E. F. Hansen, C. Rodríguez-Navarro, and K. Van Balen. Lime putties and mortars. *Studies in Conservation*, **53**(1):9–23, 2008.
11. N. E. Baker, F. R. Liewehr, T. B. Buxton, and A. P. Joyce. Antibacterial efficacy of calcium hydroxide, iodine potassium iodide, betadine, and betadine scrub with and without surfactant against e faecalis in vitro. *Oral Surgery, Oral Medicine, Oral Pathology, Oral Radiology, and Endodontology*, **98**(3):359–364, 2004.
12. R. Salazar, G. Arámbula-Villa, G. Luna-Bárcenas, J. D. Figueroa-Cárdenas, E. Azuara, and P. A. Vazquez-Landaverde. Effect of added calcium hydroxide during corn nixtamalization on acrylamide content in tortilla chips. *LWT-Food Science and Technology*, **56**(1):87–92, 2014.
13. Z. Mohammadi and P. M. H. Dummer. Properties and applications of calcium hydroxide in endodontics and dental traumatology. *International Endodontic Journal*, **44**(8):697–730, 2011.
14. United States Environmental Protection Agency. Calcium hydroxide supply chain. Executive summary. 2022. Retrieved from https://www.epa.gov/system/files/documents/2023-03/Calcium%20Hydroxide%20Supply%20Chain%20Profile.pdf.
15. Claire Jackson. Hydrated lime handling systems for thermal enhanced oil recovery: Toolkit. WaterSMART Solutions Ltd. 2016.
16. C. Williford, W.-Y. Chen, N. K. Shamas, and L. K. Wang. Lime stabilization. *Biosolids Treatment Processes*, 207–241, Humana Press, 2007.
17. T. Oates. Lime and limestone. *Kirk-Othmer Encyclopedia of Chemical Technology*, **15**:1–17, 2002.

18. M. Hassibi. An overview of lime slaking and factors that affect the process. In *Presentation to 3rd International Sorbalit Symposium*, 1–19, 1999.
19. E. E. Harper. Lime slaking. *Journal (American Water Works Association)*, **26**(6):750–756, 1934.
20. C. Rodriguez-Navarro and E. Ruiz-Agudo. Nanolimes: From synthesis to application. *Pure and Applied Chemistry*, **90**(3):523–550, 2018.
21. R. Camerini, G. Poggi, D. Chelazzi, F. Ridi, R. Giorgi, and P. Baglioni. The carbonation kinetics of calcium hydroxide nanoparticles: A boundary nucleation and growth description. *Journal of Colloid and Interface Science*, **547**:370–381, 2019.
22. J. Kariya, J. Ryu, and Y. Kato. Reaction performance of calcium hydroxide and expanded graphite composites for chemical heat storage applications. *ISIJ International*, **55**(2):457–463, 2015.
23. S. Reddy, V. Prakash, A. Subbiya, and S. Mitthra. 100 years of calcium hydroxide in dentistry: A review of literature. *Indian Journal of Forensic Medicine & Toxicology*, **14**(4):1203–1219, 2020.
24. Z. Mohammadi and P. M. H. Dummer. Properties and applications of calcium hydroxide in endodontics and dental traumatology. *International Endodontic Journal*, **44**(8):697–730, 2011.
25. R. Ba-Hattab, M. Al-Jamie, H. Aldreib, L. Alessa, and M. Alonazi. Calcium hydroxide in endodontics: An overview. *Open Journal of Stomatology*, **6**(12):274–289, 2016.
26. S. Sahebi, F. Moazami, and P. Abbott. The effects of short-term calcium hydroxide application on the strength of dentine. *Dental Traumatology*, **26**(1):43–46, 2010.
27. A. H. Gluskin, G. Lai, C. I. Peters, and O. A. Peters. The double-edged sword of calcium hydroxide in endodontics: Precautions and preventive strategies for extrusion injuries into neurovascular anatomy. *The Journal of the American Dental Association*, **151**(5):317–326, 2020.
28. S. P. Eriksen, G. M. Irwin, and J. V. Swintosky. Antacid properties of calcium, magnesium, and aluminum salts of water-insoluble aliphatic acids. *Journal of Pharmaceutical Sciences*, **52**(6):552–556, 1963.
29. R. Giorgi, L. Dei, M. Ceccato, C. Schettino, and P. Baglioni. Nanotechnologies for conservation of cultural heritage: Paper and canvas deacidification. *Langmuir*, **18**(21):8198–8203, 2002.
30. M. Ambrosi, L. Dei, R. Giorgi, C. Neto, and P. Baglioni. Colloidal particles of $Ca(OH)_2$: Properties and applications to restoration of frescoes. *Langmuir*, **17**(14):4251–4255, 2001.
31. E. Carretti, D. Chelazzi, G. Rocchigiani, P. Baglioni, G. Poggi, and L. Dei. Interactions between nanostructured calcium hydroxide and acrylate copolymers: Implications in cultural heritage conservation. *Langmuir*, **29**(31):9881–9890, 2013.
32. A. Dowling, J. O'Dwyer, and C. C. Adley. Lime in the limelight. *Journal of Cleaner Production*, **92**:13–22, 2015.
33. G. Montes-Hernandez, N. Concha-Lozano, F. Renard, and E. Quirico. Removal of oxyanions from synthetic wastewater via carbonation process of calcium hydroxide: Applied and fundamental aspects. *Journal of Hazardous Materials*, **166**(2–3):788–795, 2009.
34. Meenakshi, R. C. Fluoride in drinking water and its removal. *Journal of Hazardous materials*, **137**(1):456–463, 2006.

35. X. Wu, P. Sten, S. Engblom, P. Nowak, P. Österholm, and M. Dopson. Impact of mitigation strategies on acid sulfate soil chemistry and microbial community. *Science of the Total Environment*, **526**:215–221, 2015.

36. Other uses of lime, 2023.

37. D. E. Pettry and C. I. Rich. Modification of certain soils by calcium hydroxide stabilization. *Soil Science Society of America Journal*, **35**(5):834–838, 1971.

38. J. A. Chukwudebelu and J. Agunwamba. Potassium hydroxide and calcium hypochlorite. *SSRG International Journal of Agriculture and Environmental Science*.

39. Heathline. How is calcium hydroxide used in food, and is it safe?, 2018. Retrieved from https://www.healthline.com/health/calcium-hydroxide#pickling.

40. W. Carmen. Nixtamalization, a meso american technology to process maize at small scale with great potential for improving the nutritional quality of maize based foods. In *2nd International Workshop on Food-Based Approaches for a Healthy Nutrition*, Ouagadougou, 23–28, 2003.

41. C. Palacios, G. Cormick, G. J. Hofmeyr, M. N. Garcia-Casal, J. P. Peña-Rosas, and A. P. Betrán. Calcium-fortified foods in public health programs: Considerations for implementation. *Annals of the New York Academy of Sciences*, **1485**(1):3–21, 2021.

6 Calcium Oxide
Synthesis and Applications

Sanjeev Gautam and Ritika Charak
Advanced Functional Materials Laboratory, Dr. S.S. Bhatnagar
University Institute of Chemical Engineering and Technology
Panjab University, Chandigarh, India

6.1 INTRODUCTION

Calcium oxide(CaO) typically also known as quicklime or burnt lime, is an inorganic material with alkaline nature which is used for a wide range of applications. CaO has a CAS number of [1304-78-8] with a white to pale yellow appearance. It has a density of 3.35 g/cm^3, a melting point of 2572 °C and a boiling point of 2853 °C. It has high solubility in solutions such as glycerol, water, acids, and sugar solution and has high reactivity with water (H_2O) and carbon dioxide (CO_2) (Figure 6.1 and Table 6.1).

Due to this high reactivity, quicklime should be kept in a tightly sealed container and transported in bunkers with an adequate sealing discharge mechanism which may reduce the reactivity to water vapor and atmospheric CO_2. The chemical has a storage capacity for 5–10 days. Quicklime should be kept away from flammable materials. Any contact with water can cause a huge fire breakout in the presence of material such as wood.

The most common raw material for the production of CaO is calcium carbonate ($CaCO_3$). Calcium carbonate is extracted from limestones, which is abundant in the earth's crust. Other methods of extraction of raw materials to produce CaO from nature are by using eggshells, fish skulls, sea shells, oyster shells, etc. [3]. There are several synthesis methods for producing calcium oxide (CaO) based on different reaction processes. The most straightforward method is the thermal decomposition of calcium carbonate ($CaCO_3$) at high temperatures, where $CaCO_3$ is heated to release carbon dioxide (CO_2) and leave behind calcium oxide. Another common method is calcination of calcium hydroxide ($Ca(OH)_2$), which involves heating of calcium hydroxide to high temperatures. The sol-gel method involves mixing a precursor solution containing calcium ions with a solvent and gelation agent, followed by controlled hydrolysis and condensation reactions, drying, and calcination to obtain calcium oxide. Additionally, calcium oxide can also be synthesized through the pechini method or the polymeric precursor method. This procedure involves the

DOI: 10.1201/9781003360599-6

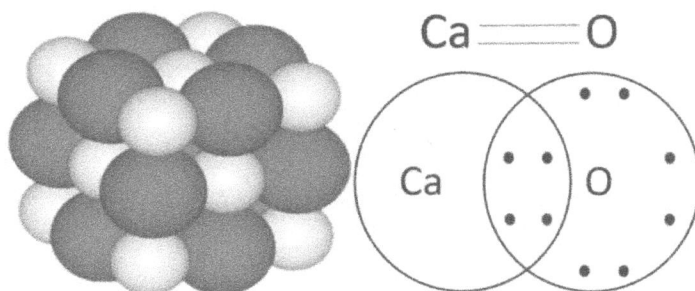

Figure 6.1 Typical unit cell and orbital structure of calcium oxide.

Table 6.1
Physical and Chemical Properties of Calcium Oxide

Structure	Cubic (Halife)
Color	White or pale yellow
Odor	Earthy odor
Molecular weight	56.0768 g per mol
Density	3.35 g/cm^3
Melting point	2572 °C
Boiling point	2583 °C
Porosity	upto 55%
Bulk density	900–1200 kg/m^3
Hardness	2–3 Mohs
CAS number	1304-78-8
Heat of hydration	1140 KJ/kg heat is released in reaction with H_2O
Std. molar entropy	40 J mol^{-1} k^{-1}
Std. enthalpy of formation	−635 KJ mol^{-1}
Affinity with water	High
Reaction with carbon dioxide	reacts above 300–800 °C
Acid neutralization	Used to dissolve acid oxides (SiO_2, Al_2O_3, Fe_2O_3)
Solubility	Soluble in acids, glycerol, sugar solution, etc.

Source: References [1, 2].

generation of a polymerizable complex by combining metal salts, chelating agents, and an organic acid together, resulting the complex to be subjected to thermal decomposition. This leads to the synthesis of the targeted metal oxide or calcium oxide in this case. The choice of synthesis method depends on factors such as the desired purity, particle size, specific surface area (SSA) and specific application requirements for the calcium oxide product. Such parameters can be analysed through techniques such as: scanning electron microscope(SEM), fourier-transform infrared spectroscopy(FTIR), X-ray diffraction (XRD), etc. shown in Figure 6.2.

Calcium oxide has been used for centuries as a binding agent in the production of mortar and cement. Due to the abundance of its raw material and its simple

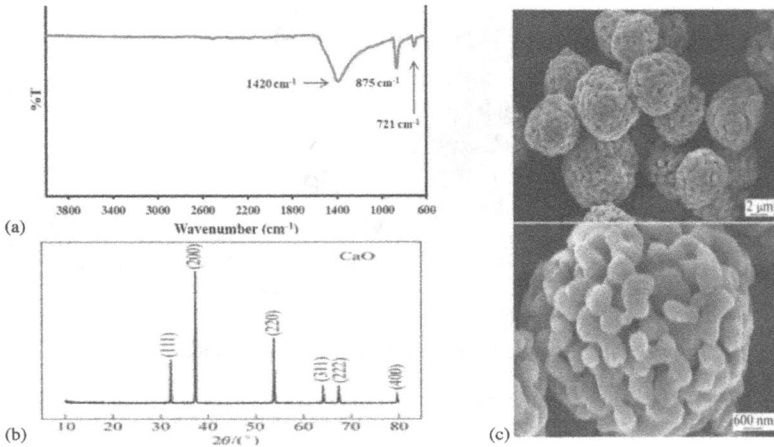

Figure 6.2 General characteristics of calcium oxide. (a) FT-IR (left top). (b) XRD (left bottom). (c) SEM images (right side) [4].

production, it is one of the oldest chemical transformation used by humankind [5]. Before the era of electric light, lime was used as a light agent in theaters. It is a highly inexpensive product which has uses in glass production and due to its reactivity with silicates, it is used in metal industry for removing impurities. Lime is employed in water and sewage treatment to mitigate acidity and serve as a flocculent in swimming pools for the removal of phosphates and other contaminants. The paper industry utilizes CaO during its production process as a coagulant and to dissolve lignin. In agriculture, lime is used to enhance acidic soils, and in gas scrubbers, it aids in the desulfurization of waste. Additionally, lime has traditional applications such as concealing the odor of decomposition in burial practices and revealing fingerprints in forensic science. Quicklime is also used in agriculture industry as a nutrient in fertilizers to help in crop production. Calcium oxide gives an exothermic reaction when mixed with water producing calcium hydroxide ($Ca(OH)_2$) which is reversible in nature. This exothermic reaction is used to make instant and portable heating packages which are used in the food industry to make self heating instant meals and as a portable heating pad for medical purposes. It is also used in portland cement (used in dental and biomedical applications) to make it cost effective [6]. Recently, calcium oxide is being used as a catalyst in the production of biodiesels, which is called transesterification. This process involves the utilization of methanol to convert triglycerides into fatty acid chains with the presence of a homogeneous or heterogeneous catalyst. With the introduction of nanomaterials, the catalytic properties are enhanced due to its higher surface area and larger grain size [7].

The consumption of lime has been increasing from last few decades. China is the largest producer of CaO at an estimate mil. tonnes per year, USA is second with estimated production of the material of about million tonnes. In 2002, the world total production represented around 120 million tonnes, whereas in 2012 the total

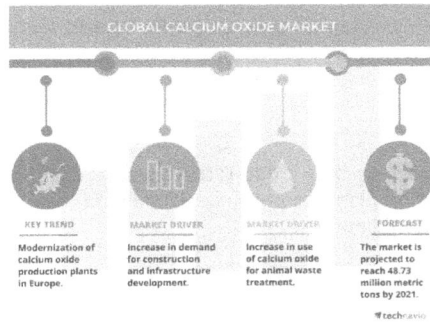

Figure 6.3 Top global markets consuming calcium oxide.

production of lime worldwide was about 340 million tonnes [8]. The market studied a decrease in production during the COVID-19 outbreak, but the market showed recovery in 2021 and 2022. The quicklime market is estimated at around 51.9 million metric tons currently(2023) with 4.86% growth rate in next five years [9] (Figure 6.3).

6.2 SYNTHESIS METHODS

The raw materials used in the synthesis of calcium oxide are commonly extracted from limestones, dolomite ores, which are widely and readily available. Other calcium-rich materials such as marble, chalk, etc can also be used in making calcium carbonate, which can then be used in production of calcium oxide. Recently, extraction for calcium carbonate has been done by organic material to recycle the waste materials and make the extraction of raw materials more environment friendly. Many scientists have used waste products and carcass of different species to produce calcium carbonate such as, Takeno et al. [3] used silver croaker fish's bones, Hussien et al. [10] used cockle shell and Gaibe et al. [11] recycled and reused eggshells for producing calcium carbonate. The various methods for synthesis of calcium oxide are given as follows. The various methods for synthesis of calcium oxide are given as follows.

6.2.1 THERMAL DECOMPOSITION OR CALCINATION

The most common way of producing calcium oxide is by thermal decomposition of limestones, which have calcium carbonate as its main component. The lime made through this process has several impurities such as: silicon oxide (SiO_2), iron oxide (Fe_2O_3), and magnesium oxide(MgO), as they are also part of the limestone mineral. The raw mineral is first crushed and sieved into smaller size and uniform shape which is then treated with very high heat. This method is called calcination (Figure 6.4),

- Calcium carbonate collected from its ores and mines is treated in the lime kiln to produce a crushed and sieved powder.
- This powder is the calcined at high temperatures to give CaO.

CaCO₃

- Produced calcium oxide is the seperated and send either to hydration plant (to produce slaked lime) or simplified as grounded lime.

CaO

- The produced CaO is then send for sales where it is tightly packed to avoid with water and inflammable materials.
- Storing of the product may vary from 5 to 10 days.

Sales

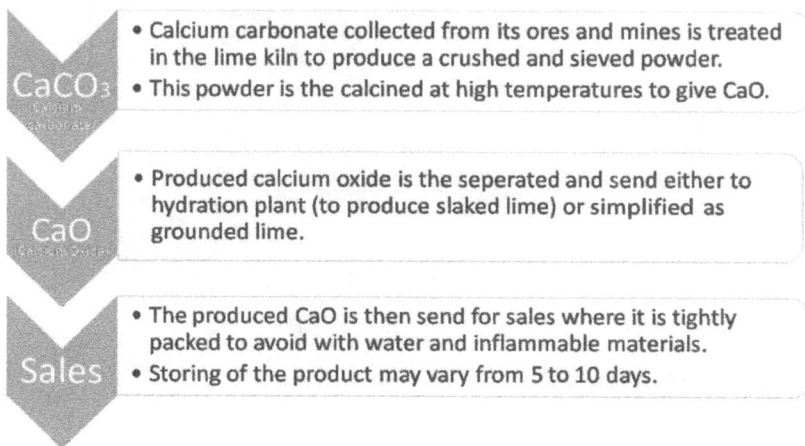

Figure 6.4 Graphical representation of production of quicklime by calcination process.

where $CaCO_3$ is heated at a high temperature of 825 °C:

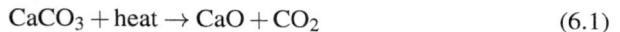

$$CaCO_3 + heat \rightarrow CaO + CO_2 \tag{6.1}$$

This reaction is reversible in nature as CaO cools down. It starts to reabsorb carbon dioxide (CO_2) from the air and form $CaCO_3$. When limestone is treated at approximately 850 °C, it yields "Standard Quicklime" which finds its use in many applications. The calcium oxide, obtained by heating $CaCO_3$ at 1250 °C, with a surface area (SSA) of about 4 cm²/g, results in dense and defect-free compound. These specimens can be fired at the range of temperatures from 1700 to 1750 °C, giving fully sintered ceramics with a density of 0.91–0.92 and a porosity of approximately 0.5%. Increasing the temperature to 2000 °C or extending the sintering period further enhances the densification, resulting in a density ranging up to 0.94. These defect free and sintered CaO, has extension to its shelf life from a few days to nearly a month [12].

This process can also happen with the calcination of calcium hydroxide ($Ca(OH)_2$). Here, the hydroxide is heated until the water content in it is evaporated. This gives CaO and water (H_2O) as its byproducts. Different calcification temperatures change the properties of the formed quicklime:

$$Ca(OH)_2 + heat \rightarrow CaO + H_2O \tag{6.2}$$

By introducing 325-mesh particles of $Ca(OH)_2$ into a hot fluidized-bed furnace at temperatures of 750–780 °C, nanoscale quicklime can be obtained. This process involves the conversion of calcium hydroxide to nano-sized calcium oxide. The resulting nanoscale calcium oxide particles possess exceptional reactivity towards air and moisture. These particles typically have sizes ranging from 20 to 80 nanometers and a SSA between 15,000–50,000 square meters per kilogram. They are available in various forms, including ultrahigh and high purity forms, as well as transparent, coated, dispersed particles, and even as a nanofluid [12].

6.2.2 SOL-GEL METHOD

The sol-gel method uses low pressure and temperature for the preparation of calcium oxide. This makes it an ideal method for production of quicklime in a cost effective manner. The technique comprises dissolving the molecular precursor in either water or alcohol, which is the converted into gel by heating the solution and then stirred by hydrolysis or alcoholysis. Next, the gel is dried using methods according to the desired properties that the material should acquire. The dried gel is powdered and then calcined for further characterization. Here, the limestone is first crushed and then washed, dried, grounded and sieved. The prepared powder is then dissolved in 1M hydrochloric acid (HCl) producing calcium chloride ($CaCl_2$), water and carbon dioxide as its end products [13].

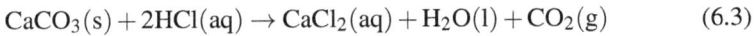

$$CaCO_3(s) + 2HCl(aq) \rightarrow CaCl_2(aq) + H_2O(l) + CO_2(g) \qquad (6.3)$$

For the formation of sol we use a hydrolysis process. A solution of sodium hydroxide (NaOH) and calcium chloride is made by slowly pouring 1M of NaOH into $CaCl_2$. This converts the solution into a sol at a standard temperature. Due to pouring NaOH slowly, subsequent precipitation of sodium hydroxide takes place, resulting the small particles to interconnect with each other and forms a highly crystalline and rigid gel. The gel is then aged for a day at a constant temperature.

$$CaCl_2(aq) + 2NaOH(aq) \rightarrow Ca(OH)_2(s) + 2NaCl(aq) \qquad (6.4)$$

The aged gel is cleaned with distilled water, thus removing impurities in the precipitate and the gel is dried for another day in the oven at 60 °C and calcined at 900 °C for one hour which completes the sol-gel synthesis process (Figure 6.5).

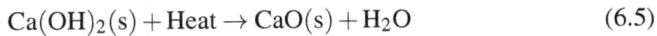

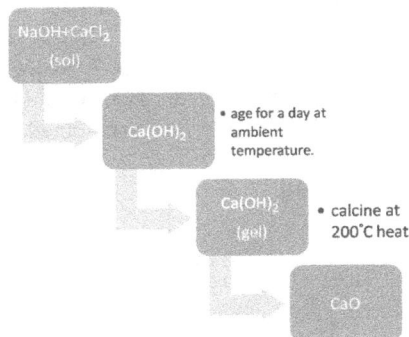

$$Ca(OH)_2(s) + Heat \rightarrow CaO(s) + H_2O \qquad (6.5)$$

Figure 6.5 Graphical representation of synthesis of quicklime by sol-gel method.

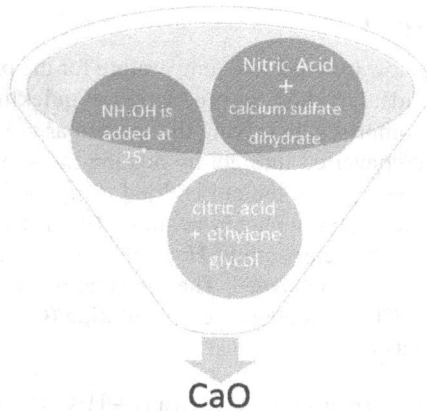

Figure 6.6 Graphical representation of synthesis of quicklime by pechini method.

6.2.3 PECHINI PROCESS FOR SYNTHESIS OF CALCIUM OXIDE

The Pechini method (Figure 6.6), commonly known as the polymeric precursor method is a versatile technique used for synthesis of complex metal oxides. This procedure involves the generation of a polymerizable complex by combining metal salts, chelating agents, and an organic acid. The resulting complex is then subjected to thermal decomposition, leading to the synthesis of the targeted metal oxide product. For obtaining calcium oxide, 10.0305 ml of ethylene glycol is heated to 70 °C, and 9.4559 g of citric acid monohydrate is slowly added, at a constant temperature of 70 °C. The mixture is continuously shaken until a transparent solution is obtained. Simultaneously, a solution of the precursor is prepared with 0.3M and 2.5356g of calcium sulfate dihydrate ($CaSO_4 \cdot 2H_2O$) in an aqueous solution of 0.4M of nitric acid (HNO_3). The precursor solution prepared is then added to the ethylene glycol and citric acid mixture and when it became semitransparent, the temperature is reduced to 25 °C. The temperature is lowered to add NH_4OH until a solution at pH 5 was obtained. The transparent system is heated at 90 °C with constant shaking until a white resin is formed. The resin is pre-calcined at 300 °C and the soaked and crushed using an agate mortar. The resulting powder is then calcined at 800–900 °C [14].

6.3 APPLICATIONS

Calcium oxide is an important and significant (Figure 6.7) chemical that is used in a wide variety of industries, which covers a wide range of applications. CaO is easily produced and is a recyclable material, which makes a cost effective and environment friendly material. Various applications of calcium oxide and its significance in these industries are listed as follows.

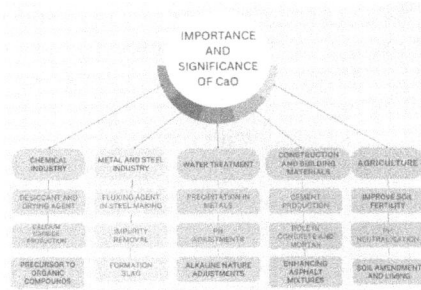

Figure 6.7   Importance and significance of calcium oxide.

6.3.1   BUILDING AND CONSTRUCTION INDUSTRY

One of the oldest use of quicklime has been in the building and construction industry. Dating back to the Roman era, limestones have been crushed and calcined into oxides and hydroxides of calcium which have been used as a binder, cement and mortar in the industry [15, 16].

Aerated Concrete: Ground quicklime is combined with pulverized fuel ash or ground silica, water, aluminum powder, sand, and cement of controlled reactivity to create aerated concrete blocks. The reaction between quicklime and aluminum powder produces hydrogen bubbles, resulting in the expansion of the mixture. Simultaneously, the quicklime reacts with water, generating heat and facilitating the setting of the mixture. Once set, the mixture is extracted from the mold and cut into blocks. Finally, the blocks undergo treatment in steam sterilizers, such as an autoclave and then are subjected to high temperature and pressure [2].

Soil Stabilization and Pavement Treatment: Quicklime and hydrated lime can significantly enhance the load-carrying capacity of soils containing clay. This is achieved through a chemical reaction with the finely divided silica and alumina present in the soil, which results in the formation of calcium silicates and aluminates that exhibit cementing properties. Quicklime offers an advantage over hydrated lime in terms of soil drying as it makes increases the capacity of the soil to absorb up to 30% of its weight in moisture and generates heat which helps in accelerating the evaporation process. Soil stabilization is commonly employed in road and railway construction to strengthen the subgrade, thereby reducing the required construction depths. Quicklime can also be used to create a sub-base instead of relying on aggregate materials. In construction sites characterized by high clay content, soil stabilization facilitates the placement and compaction of on-site materials. Lime piling technique which consist of digging deep holes which are filled with quicklime, is occasionally employed in certain countries to stabilize soft soils and provide support for building slabs and failing slopes [17].

Asphalt Mixture: Hydrated lime serves a crucial role as an anti-stripping agent in hot mix asphalt. This helps in prevention of loss in adhesion between the aggregate surface and the asphalt cement binder in the presence of moisture. It is also used in cold in-place recycling to rehabilitate distressed asphalt pavement. Research has demonstrated the effectiveness of hydrated lime in reducing stripping in hot mix asphalt, acting as a mineral filler to enhance the stiffness of the asphalt binder [18]. Additionally, hydrated lime improves resistance to low-temperature cracking, modifies oxidation kinetics, and interacts with oxidation by-products to mitigate their detrimental effects. It also modifies the plastic properties of clay fines, enhancing moisture stability and durability. Incorporating lime in hot asphalt mixtures leads to significant cost savings, with studies reporting savings exceeding 45% and a reduction of approximately $20 per ton. Moreover, life cycle studies indicate an increase in pavement lifespan of approximately 38%. There have also been reports of a method involving the use of lime pellets for manufacturing asphalt [19].

6.3.2 PURIFICATION OF WATER

Conventional water treatment encompasses a series of unitary processes: pretreatment, coagulation, flocculation, sedimentation, filtration, and disinfection.

Pretreatment includes, the removal of settleable materials, pre-oxidation, adsorption, and pre-alkalization. Pre-alkalization is a necessary process when the alkalinity of the raw water is not enough for achieving optimal coagulation. Thus, the raw water is conditioned with the addition of chemicals like soda ash (Na_2CO_3), caustic soda (sodium hydroxide, NaOH), or hydrated lime ($Ca(OH)_2$). During coagulation, the alkalinity reacts with aluminum sulfate, resulting in a decrease in alkalinity accompanied by a decrease in pH. Coagulation involves the neutralization of turbidity-causing particles via the introduction of a coagulant (e.g., aluminum sulfate), facilitating their aggregation into larger and denser entities known as flocs. This process, referred to as flocculation, necessitates controlled conditions, as excessive agitation can disrupt formed flocs.

Sedimentation or decantation represents the initial and crucial step in effectively separating particles from water. Filtration, is the water clarification process, it is also regarded as a key barrier for the retention of pathogenic microorganisms. Additionally, if the pH decrease resulting from coagulation falls below the specified quality specifications, the addition of an alkaline agent may be necessary during this stage. The output of the filtration process is termed filtered water. Disinfection serves as the final stage in the water purification process and entails the introduction of chemical agents to eliminate pathogenic microorganisms that may pose health risks. Strict adherence to turbidity and color parameters is crucial since deviations can impede the efficacy of disinfectants [20].

The lime and soda ash method is a significant process for softening raw water. Lime and soda ash are introduced to the raw water either in dry form gravimetrically or in milk of lime or solution volumetrically. In cases where storage is a challenge, hydrated lime is utilized, while quicklime (CaO) is preferred to larger plants due to its cost-effectiveness. These substances are typically mixed, agitated, and settled, following a similar approach to the purification treatment [21].

6.3.3 STEEL AND METAL INDUSTRY

Quicklime, when finely grounded is used in small quantities in fines to produce iron agglomerates. By adding approximately 6% of lime to the ore, a significant increase in the production capacity of the sinter strand is achieved. Ground quicklime also plays a crucial role in the desulfurization process of iron before it is charged into the steelmaking furnace. However, the primary application of quicklime lies in the basic oxygen steelmaking process. This entails using approximately 35–70 kg of screened quicklime per ton of steel production. The main purpose of quicklime in this process is to neutralize acidic oxides such as SiO_2, Al_2O_3, and Fe_2O_3, ultimately leading to the production of a typical molten slag [22]. The formation of the slag is essential for effective refining. For further insights into the role of quicklime and slag in removing impurities like phosphorus and sulfur from steel, extensive literature is available.

Additionally, in the basic oxygen steel making process vessel, calcined dolomite is added to achieve a slag composition containing approximately 6% MgO. This serves to reduce slag viscosity and protect the basic refractory lining, thereby enhancing the overall efficiency of the process.

Furthermore, quicklime finds application in the electric arc steelmaking process where it reacts with acidic oxides, contributing to the generation of a molten slag. Screened and ground quicklime is also widely employed in various secondary steel-making processes. For instance, the argon oxygen decarburization process relies on screened quicklime with a low residual $CaCO_3$ level (typically below). Moreover, researchers have reported the use of lime-based additives for steel smelting, show-casing the continuous exploration of innovative approaches in the steel industry.

6.3.4 PAPER AND GLASS INDUSTRY

Paper Production: During the process of paper making, all the derivatives of lime are used in different stages of the process. Concentrated sodium hydroxide (NaOH) and sodium sulphide (Na_2S) helps in breaking down the wood chips by degrading the lignin macromolecules that bind them together in the Kraft cycle process. During the digestion of wood chips, lignin is broken down, resulting in the release of organic acids and resinous compounds. These byproducts are neutralized by sodium hydroxide, which converts to the acids and resinous compound to sodium carbonate (Na_2CO_3) which is an alkaline salt [23].

After digestion of the wood chips, the residual liquor also known as 'black liquor' is separated from the pulp. It consists of both organic and inorganic compounds. The separated Black Liquor is then concentrated to reduce the water content. The organic compounds are then fed into a controlled incineration process, with the resulting organic compounds to be carried away by fuel gases. The inorganic fraction is re-covered as a smelt containing sodium carbonate and sodium sulphide. The smelt is then rehydrated, transforming it into the 'Green Liquor'.

In the Kraft cycle, the sodium sulphide is retained (22–23 g per L) and the sodium carbonate in Green Liquor is separated and converted into a stronger alkali NaOH solution (60–90 g/L). For this conversion of Na_2CO_3 to NaOH, hydrated lime $(Ca(OH)_2)$ is added to the weak alkali, which reacts with it and produces NaOH and calcium carbonate which is also known as calcium mud $(CaCO_3)$. The lime mud is then segregated through the sedimentation process. The separated calcium carbonate $(CaCO_3)$ is burned in a lime kiln to produce quicklime (CaO) for reuse. Additional lime is added to compensate for losses of sodium carbonate (Na_2CO_3) in the Kraft cycle also known as the dead load, which cannot be recycled after the process. Quicklime plays a crucial role in the process in mitigating energy costs and emissions associated with the lime kiln. The presence of deadload Na_2CO_3 in the Kraft cycle also incurs additional energy costs, estimated to be around 1 MJ per kg of NaOH is regenerated, in the form of additional dilution water that requires more pumping and heating throughout the paper making process [24].

Glass Production: The soda-lime-silica system is widely utilized in the production of various glass products. The common composition of the glass is given as $(25-x)Na_2O(x)CaO(75)SiO_2$ were examined [25]. These compositions maintained a constant silica content, while CaO replaced Na_2O, keeping the number of non-bridging oxygens (NBO) theoretically unchanged. These glass compositions are commonly used in float glass manufacturing.

According to the research done by Cormack et al., the MD simulations employed a potential model based on BKS interactions for Si-O and O-O, along with empirically derived interactions for Ca-O and Na-O. The coordination number of calcium was found to be approximately six, exhibiting a more favorable first coordination shell compared to sodium. NBOs exhibited a preference for bonding with calcium, suggesting that calcium plays a similar modifying role as sodium in the glass structure. Upon substitution, calcium replaced sodium and occupied regions rich in modifiers within the glass structure [26]. Importantly, the distribution of ring sizes in the glass remained unaltered during the substitution process.

6.3.5 CHEMICAL INDUSTRY

As a precursor in production of calcium carbide and hydrogen: It involves the reaction of calcium oxide (lime) with carbonaceous materials, typically petroleum coke or coal, in an electric arc furnace. The intense heat generated in the furnace causes the carbon in the coke or coal to reduce the calcium oxide, forming calcium carbide (CaC_2). The reaction can be represented as follows:

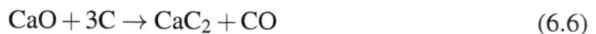

$$CaO + 3C \rightarrow CaC_2 + CO \tag{6.6}$$

Hydrogen, a versatile "green energy" carrier and co-reagent in the chemical industry which can also be produced by calcium carbide residue (CCR), a byproduct of acetylene production. By mixing CCR with various metals and heating it to 500–700 °C, hydrogen evolution was observed, with yields reaching up to 99%. This use of waste CCR offers a dual benefit of hydrogen generation and reuse in carbide synthesis. Additionally, the researchers demonstrated the production of deuterated CCR by hydrolyzing calcium carbide with deuterium oxide, followed by sintering with calcium metal to release D2-gas. This deuterated hydrogen was then employed in the synthesis of D-labeled compounds, achieving up to 75% deuterium incorporation [27].

Lime cycle: Lime cycle refers to the cycle calcium containing minerals interact with water and carbon dioxide to make new calcium compounds. The lime cycle initiates with the mineral calcium carbonate ($CaCO_3$), which follows up with chemical conversions involving intermediary compounds as quicklime (calcium oxide, CaO) and slaked lime (calcium hydroxide, $Ca(OH)_2$) which eventually regenerates to calcium carbonate. This knowledge about the lime cycle has been used in the medieval times as a great influence in changing construction practices [28]. The cycle's starts with calcium carbonate being subjected to thermal decomposition, or calcination, under elevated temperatures which yields calcium oxide. The reaction of this freshly formed oxide with water gives an exothermic reaction which results in the formation of calcium hydroxide also known as slaked lime. The resultant slaked lime is reversed through a carbonation reaction, wherein it reacts with carbon dioxide and ultimately forms calcium carbonate. This completes the lime cycle. The products created from repeated cycles may differentiate its characterstics to the original product. Grasa et al. [29] investigates the reactivity of highly cycled CaO to observe changes in the structure as well as reactivity and compares it with the original product.

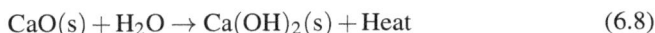

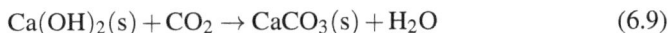

$$CaCO_3 + heat \rightarrow CaO + CO_2 \tag{6.7}$$

$$CaO(s) + H_2O \rightarrow Ca(OH)_2(s) + Heat \tag{6.8}$$

$$Ca(OH)_2(s) + CO_2 \rightarrow CaCO_3(s) + H_2O \tag{6.9}$$

During the Kraft cycle, the calcination/carbonation loop or the lime cycle (Figure 6.8) is used to recycle the raw materials back to its original form and also it compensates for the waste products to make the process cost effective and eco-friendly. This cycle plays a significant role in regulating the Earth's carbon dioxide levels and is an essential component of the carbon cycle. Abanades et al. [30] in his discusses about the significance of CaO as an absorbant which can remove carbon dioxide (CO_2) through carbonation and calcination loop. This creates a highly efficient carbon capture system at a very low cost.

Use as catalyst: As a recent application, CaO has been in use as a catalyst in making biofuel. CaO nanoparticles and microspheres are being used to develop simple and low cost catalyst which can have potential applications fields due to its well-defined morphology and simple synthetic route [4]. Transesterification is the conversion of biodegradable organic compounds such as animal fat and vegetable to biofuels. It is the chemical process of conversion of triglycerides with alcohol to alkyl esters in the presence of a heterogenous catalyst. Since homogenous catalyst are not regenerative, heterogenous catalyst or biocatalyst such as calcium oxide is an optimal material for biodiesel synthesis. Equivalent catalytic activity to that of homogeneous alkaline, catalyst stability, and the Ca^{2+} leaching makes CaO catalyst less efficient for biodiesel production. To mitigate these problems, CaO both in pure and as a mixture with other metals or their oxides, is used as a heterogeneous alkaline catalyst [11,31].

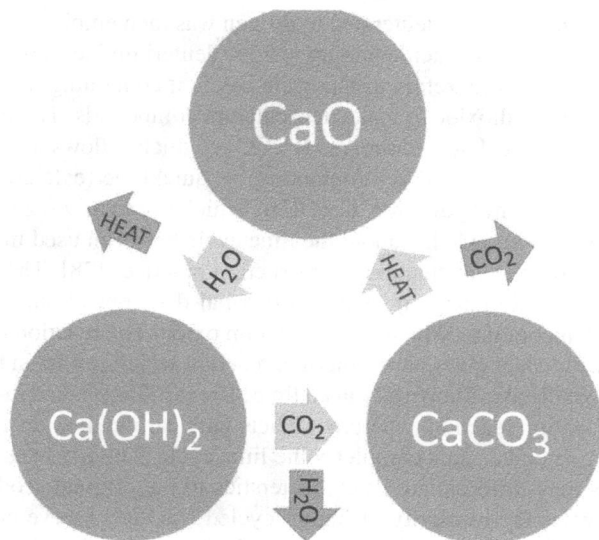

Figure 6.8   Representation of lime cycle.

Calcium oxide nanoparticles are also used as a photocatalyst in the degradation of dyes from textile industry. CaO nanoparticles can collect pollutants close to the surface of the catalyst and then the degradation process occurs. Moreover, fouling on the catalyst can be mitigated by addition of photocatalyst, which will increase the wetting properties of the catalyst membrane [32, 33].

6.3.6 AGRICULTURAL INDUSTRY

In Making Sustainable Fertilizers: Common elements such as potassium, nitrogen, and phosphorus are essential for improving plant growth. Potassium, in particular, increases the growth of a crop by 20% in soils low in potassium levels. This helps the crop in improved resistance to drought, cold, and pathogens. However, the majority of global potassium fertilizer relies on traditional water-soluble potassium salts, which represent less than 1% of global potassium reserves. To address this limitation, researchers have explored the comprehensive utilization of K-feldspar ore.

Li et al. [34] did a study, where K-feldspar was mixed with limestone ($CaCO_3$) and dolomite ($CaMg(CO_3)_2$) as additives, then subjected to high-temperature calcination. This process produced Si-Ca-K-Mg fertilizer, improves crop yield by providing essential elements for the plant. Additionally, it acts as a soil conditioner, correcting soil acidity and immobilizing harmful metals such as cadmium and chromium. The Si-Ca-K-Mg fertilizer exhibits slow-release characteristics, making it well-suited for soils in a tropical climate which may have low cation exchange capacity.

During calcination, the raw materials undergo transformations, resulting in the formation of gehlenite ($Ca_2Al_2SiO_7$) and akermanite ($Ca_2MgSi_2O_7$). Gehlenite and akermanite are formed at different temperatures, so there are occurrence of two separate reactions as:

$$Al_2O_3 + SiO_2 + 2CaO \rightarrow Ca_2Al_2SiO_7 \qquad (6.10)$$

$$MgO + 2SiO_2 + 2CaO \rightarrow Ca_2MgSi_2O_7 \qquad (6.11)$$

At higher temperatures, new phases such as rankinite ($Ca_3Si_2O_7$) appear, while crystallinity decreases, and amorphous substances are observed. The resulting fertilizer contains approximately 6.97 wt% K_2O, 42.03 wt% SiO_2, 32.86 wt% CaO, and 4.09 wt% MgO.

In conclusion, Si-Ca-K-Mg fertilizer offers a sustainable approach for efficient nutrient utilization and soil improvement in agricultural practices [34].

Sugar Refining: During the refining process of sugar beet, the initial sugar solution undergoes treatment with milk of lime to form calcium salts of organic and phosphoric acids. The solution is filtered after which it is neutralized using carbon dioxide, leading to the precipitation of calcium carbonate. The filtration process eliminates the solids which results in a purified sugar solution. Given the substantial quantities of lime (250 kg per ton of sugar) and carbon dioxide required, beet sugar producers typically incorporate lime kilns as an essential component of their operations. In contrast, refining raw sugar obtained from sugar cane demands significantly less hydrated lime (typically 5 kg per ton) due to its higher purity compared to sugar beet [35].

6.4 HANDLING AND PRECAUTIONS

Quicklime exhibits strong alkalinity when it is exposed to water or to a highly humid environment, resulting in a high pH 12.4. However, it is important to note that the substance can pose occupational health hazards in the form of airborne dust. In the United Kingdom, the occupational exposure standards (OES) for lime dust are set at 2 mg/m^3 for quicklime. Inhalation of these dust particles will severely irritate the respiratory system and can lead to subsequent inflammation. Furthermore, direct exposure of quicklime to the eyes can cause significant discomfort, redness and irritation which will require immediate treatment to prevent potential severe damage [36]. Quicklime is classified as irritants and has the potential to induce "chemical burns" on the skin, particularly when there is moisture or perspiration present, along with skin abrasion. According to the the the EEC Dangerous Preparations Directive,the product is currently categorized under risk phrases R38 (irritating to the skin) and R41 (risk of serious damage to the eyes). Prolonged and repeated skin contact with lime can cause dryness, cracking, and even dermatitis. Ingestion in small amounts of quicklime can also lead to severe damage and potential corrosion in the digestive tract. Quicklime finds application in the treating of drinking water, where regulatory limits are strictly imposed on the content of minor and trace elements [37].

6.4.1   ENVIRONMENTAL IMPACT

The extraction of limestone (primary raw material for CaO) is a tremendous noise and air pollutant.The blasting and stone processing can disturb nearby residents and structures [38].

To mitigate these impacts, good operating practices and control measures are recommended. Air pollution is another issue, as quicklime production is subject to regulations aiming to reduce emissions of pollutants such as hydrochloric acid, mercury, particulate matter, and total hydrocarbons. Proposed revisions to cement air toxic rules may lead to even stricter emissions requirements for the lime industry. Dust emissions from stone processing, lime kilns, and solids handling are also significant. Various techniques, including dust suppression and the use of bag filters or wet scrubbers, are employed to control dust and minimize environmental impacts [2]. Sulfur dioxide emissions from lime kilns are low because of the capture and retention of sulfur by quicklime, but in some processes, emissions may need to be abated by using low-sulfur fuels. Oxides of nitrogen emissions vary depending on the type of lime being produced, and it may consider techniques used in other industries for reducing nitrogen oxide emissions.

Other substances, such as volatile organic compounds, dioxins, furans, and heavy metals, typically have negligible emissions from lime kilns. Lastly, energy usage is a significant aspect of quicklime production, with kilns consuming a large portion of the total energy as fuel. Efforts are made to improve energy efficiency while maintaining product quality. Overall, addressing these environmental issues is crucial for sustainable quicklime production. Other safety measures in the lime industry encompass appropriate safeguards for moving machinery, effective electrical isolation during maintenance procedures, obligatory use of eye protection in lime plants, and provision of eyewash facilities to ensure prompt and appropriate first aid.

6.5   CONCLUSION

The abundance of lime in the earth's crust makes it readily available, cheap and easy to synthesize, which makes calcium oxide, a versatile material, maintaining its relevance throughout the history of mankind. Calcium oxide is the global consumption of lime is increasing every year, it is being used in making various building cement, mortars, water purification, glass manufacture, etc. CaO is widely in being the precursor in the production of materials such as calcium hydroxide, calcium carbide which are very useful derivative of calcium oxide Quicklime has found its recent use as a catalyst in the process of synthesis of bio-fuels. While being slow in catalysis than many catalysts, the material is doped to make it stable for longer durations and catalysis. The calcium carbide cycle, which includes quicklime as the part of its cycle is used in cost effective production of hydrogen, which is one of the green energy carriers of the chemical industry. From recent researches, one can conclude that CaO has become an important part for the sustainable development initiative where it has also been used in fertilizers to develop more stable and ecological fertilizer. While the product in these forms is an advantage but its production and extraction are still huge noise and air pollutants. Keeping this in mind several steps are taken to decrease pollution from CaO production as the raw materials are substituted with natural inorganic sources, such as sea shells, fish heads, egg shells, etc. Safety measures are taken in the lime industry to decrease high energy consumption, dust emission and emission of volatile compounds. This makes the extraction to consumption of CaO to be done in a sustainable manner, making it a diverse and multi-purpose calcium-based material.

REFERENCES

1. S. Zumdahl. Chloride (cl). Chemical principles 7th edition. Retrieved from *http://www. m.webmd.com/atozguides/chloride-cl*, 2009.
2. Tony Oates. Lime and limestone. *Kirk-Othmer encyclopedia of chemical technology*, 1–53, 2000.
3. M. L. Takeno, I. M. Mendonça, S. de S. Barros, P. J. de Sousa Maia, W. A. G. Pessoa Jr., M. P. Souza, E. R. Soares, R. dos S. Bindá, F. L. Calderaro, I. S. C. Sá, C. C. Silva, L. Manzato, S. Iglauer, F. A. de Freitas. A novel CaO-based catalyst obtained from silver croaker (Plagioscion squamosissimus) stone for biodiesel synthesis: Waste valorization and process optimization. *Renewable Energy*, **172**:1035–1045, 2021.
4. H.-X. Bai, X.-Z. Shen, X.-H. Liu, and S.-Y. Liu. Synthesis of porous CaO microsphere and its application in catalyzing transesterification reaction for biodiesel. *Transactions of Nonferrous Metals Society of China*, **19**:s674–s677, 2009.
5. S. Holmes and M. Wingate. Building with lime. *A practical introduction*, 320, 1997.
6. A.S. Wagh. Calcium phosphate cements: Chemically bonded phosphate ceramics. *twenty-first century materials with divese applications*, pages 17–34, 2016.
7. J. Jayaprabakar, A. Karthikeyan, K. V. Anand, T. Arunkumar, N. Anbazhaghan, and G. Rangasamy. Synthesis and characterization of calcium oxide nano particles obtained from biowaste and its combustion characteristics in a biodiesel operated compression ignition engine. *Fuel*, **350**:128839, 2023.
8. A. Miskufova, T. Havlik, B. Bitschnau, A. Kielski, and H. Pomadowski. Properties of CaO sintered with addition of active alumina. *Ceramics–Silikáty*, **59**(2):115–124, 2015.
9. Mordor Intellegence. Calcium oxide market analysis, 2023. Retrieved from https://www. mordorintelligence.com/industry-reports/calcium-oxide-market.
10. A. I. Hussein, Z. Ab-Ghani, A. Nazeer C. Mat, N. A. Ab Ghani, A. Husein, and I. Ab. Rahman. Synthesis and characterization of spherical calcium carbonate nanoparticles derived from cockle shells. *Applied Sciences*, **10**(20):7170, 2020.
11. I. Gaide, V. Makareviciene, E. Sendzikiene, and M. Gumbytė. Rapeseed oil transesterification using 1-butanol and eggshell as a catalyst. *Catalysts*, **13**(2):302, 2023.
12. R. C. Ropp. Group 16 (O, S, Se, Te) alkaline earth compounds. *Encyclopedia of the alkaline earth compounds*, 105–197, Elsevier, 2013.
13. D. Bokov, A. T. Jalil, S. Chupradit, W. Suksatan, M. J. Ansari, I. H. Shewael, G. H. Valiev, and E. Kianfar. Nanomaterial by sol-gel method: Synthesis and application. *Advances in Materials Science and Engineering*, **2021**:1–21, 2021.
14. M. R-Joya, A. M. Raba, J. J. Barba-Ortega. Synthesis of calcium oxide by means of two different chemical processes. *Universidad, Ciencia y Tecnología*, **20**(81):188–192, 2016.
15. N. J. Delatte. Lessons from roman cement and concrete. *Journal of Professional Issues in Engineering Education and Practice*, **127**(3):109–115, 2001.

16. A. Moropoulou, A. Bakolas, and E. Aggelakopoulou. The effects of limestone characteristics and calcination temperature to the reactivity of the quicklime. *Cement and concrete Research*, **31**(4):633–639, 2001.

17. C. D. F. Rogers and S. Glendinning. Improvement of clay soils in situ using lime piles in the UK. *Engineering Geology*, **47**(3):243–257, 1997.

18. P. E. Sebaaly, D. N. Little, and J. A. Epps. The benefits of hydrated lime in hot mix asphalt. Technical report, University of Neveda, Reno, 2006.

19. N. R. Hill, S. Holmes, and D. Mather. *Lime and other alternative cements*. Practical Action Publishing, 1992.

20. E. Mena and F. Piñeiro. Use of quicklime in drinking water treatment. *Engineering research*. **2**:13, 2023.

21. D. S. Arnold. *Water Purification Today*. Ohio State University, College of Engineering, **27**(2):7–8, 20, 28, 1943.

22. A. Jackson. Oxygen steelmaking for steelmakers. *BUTTERWORTH AND CO. LTD., LONDON. 1969, 358 P*, 1969.

23. A. Dowling, J. O'Dwyer, and C. C. Adley. Lime in the limelight. *Journal of Cleaner Production*, **92**:13–22, 2015.

24. T. T. Chen and S. S. B. Wang. Constituents of calcined limestone and their relevance in paper manufacturing. *The Canadian Mineralogist*, **24**(2):303–306, 1986.

25. M. P. Syomov and O. V. Rakhimova. Glasses. In *Encyclopedia of Analytical Science: Second Edition*, 208–214. 2005.

26. A. N. Cormack and J. Du. Molecular dynamics simulations of soda–lime–silicate glasses. *Journal of Non-Crystalline Solids*, **293**:283–289, 2001.

27. K. A. Lotsman and K. S. Rodygin. Calcium carbide residue–a promising hidden source of hydrogen. *Green Chemistry*, **25**(9):3524–3532, 2023.

28. K. Van Balen. Understanding the lime cycle and its influence on historical construction practice. In *Proceedings of the First International Congress on Construction History*, volume 20, 41–54. Instituto Juan de Herrera Universidad Politécnica de Madrid, 2003.

29. G. S. Grasa, J. C. Abanades, M. Alonso, and B. González. Reactivity of highly cycled particles of CaO in a carbonation/calcination loop. *Chemical Engineering Journal*, **137**(3):561–567, 2008.

30. J. C. Abanades, E. J. Anthony, J. Wang, and J. E. Oakey. Fluidized bed combustion systems integrating CO_2 capture with CaO. *Environmental Science & Technology*, **39**(8):2861–2866, 2005.

31. H. Li, Y. Wang, X. Ma, Z. Wu, P. Cui, W. Lu, F. Liu, H. Chu, and Y. Wang. A novel magnetic CaO-based catalyst synthesis and characterization: Enhancing the catalytic activity and stability of cao for biodiesel production. *Chemical Engineering Journal*, **391**:123549, 2020.

32. A. Anantharaman, S. Ramalakshmi, and M. George. Green synthesis of calcium oxide nanoparticles and its applications. *International Journal of Engineering Research and Application*, **6**(10):27–31, 2016.

33. A. B. D. Nandiyanto, B. S. Maharani, and R. Ragaditha. Calcium oxide nanoparticle production and its application as photocatalyst. *Journal of Advanced Research in Applied Sciences and Engineering Technology*, **30**(3):168–181, 2023.

34. X.-Y. Li, J. Long, P.-Q. Peng, Q. Chen, X. Dong, K. Jiang, H.-B. Hou, and B.- H. Liao. Evaluation of calcium oxide of quicklime and Si–Sa–Mg fertilizer for remediation of Cd uptake in rice plants and Cd mobilization in two typical Cd-polluted paddy soils. *International Journal of Environmental Research*, **12**:877–885, 2018.

35. F. A. Mostafa, A. N. Gad, A.-A. M. Gaber, and A.-M. A. Abdel-Wahab. Preparation, characterization and application of calcium oxide nanoparticles from waste carbonation mud in clarification of raw sugar melt. *Sugar Tech*, **25**(2):331–338, 2023.

36. W. S. Cain, A. A. Jalowayski, M. Kleinman, N. S. Lee, B. R. Lee, B. H. Ahn, K. Magruder, R. Schmidt, B. K. Hillen, C. B. Warren, B. D. Culver. Sensory and associated reactions to mineral dusts: Sodium borate, calcium oxide, and calcium sulfate. *Journal of Occupational and Environmental Hygiene*, **1**(4):222–236, 2004.

37. G. D. Wadelin. The applicability and use of environmental assessment methodologies for industrial processes. The University of Manchester (United Kingdom), 2004.

38. D. M. Hiller and G. I. Crabb. Groundborne Vibration Caused by Mechanised Construction Works. Transport Research Laboratory, 7–12, 2000.

7 Environment Application of Natural Materials

Ruhani Baweja and Sanjeev Gautam
Advanced Functional Materials Laboratory, Dr. S.S. Bhatnagar
University Institute of Chemical Engineering and Technology
Panjab University, Chandigarh, India

7.1 INTRODUCTION

Natural materials, derived from plants, minerals, and animals, have been utilized for centuries by various civilizations because of their diverse properties and applications. However, with the increase in technological development, there has been a significant rise in the usage of artificial materials. Unfortunately, this increased usage of artificial materials has had hazardous effects on both our health and the environment [1]. The three main pillars of sustainability are the economic, environmental, and social aspects. Among these, protecting the environment and persuading sustainability have become major interests in society. Not only do they contribute to a smoother living, but they also prolong the life of human beings and preserve resources for future generations [2]. There is a renewed interest in exploring sustainable alternatives to conventional materials.

Throughout history, humans have relied on the bounties of nature to meet their needs, constructing shelters from wood, creating clothing from natural fibers, and using medicinal plants for healing. These practices not only exemplify our deep connection with the natural world but also highlight the potential of natural materials in sustainable living. However, with the advent of the industrial revolution and modern technological advancements, there was a significant shift towards synthetic materials derived from fossil fuels. While these materials offered new possibilities and convenience, they also brought about a range of environmental problems [3]. For instance, plastics revolutionized packaging, but their durability and resistance to degradation led to widespread plastic pollution, contaminating oceans, soils, and even entering the food chain [4]. Figure 7.1 describes many sources of plastic pollution. Plastic pollution is rampant in various parts of the world, exploiting the environment rapidly. The increased usage of plastics, especially microplastics, is causing hazardous effects on fish and humans alike [5]. The accumulation of plastic waste in the world's oceans, in particular, has garnered global attention. Large floating patches of plastic debris, such as the Great Pacific Garbage Patch, have a detrimental impact on marine

DOI: 10.1201/9781003360599-7

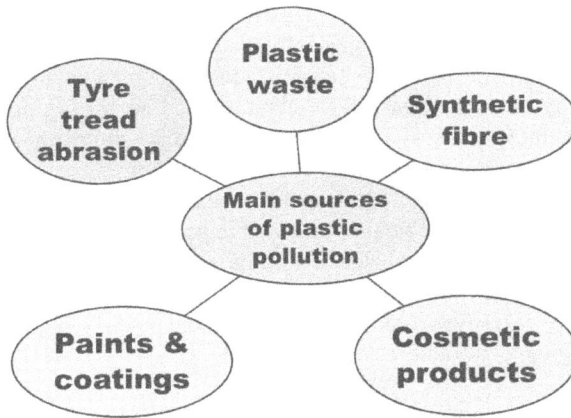

Figure 7.1 Causes of plastic pollution.

life, disrupting ecosystems, and causing harm to aquatic animals through ingestion and entanglement [6]. As a result, many regions are turning towards more environmentally friendly products, such as natural fibers [7, 8], which can serve as better substitutes for these plastics.

The burning of fossil fuels for energy production, transportation, and industrial processes releases vast amounts of greenhouse gases, such as carbon dioxide (CO_2), methane (CH_4), and nitrous oxide (N_2O), into the atmosphere. These gases trap heat and contribute to the greenhouse effect, leading to global warming and climate change [9]. The consequences of climate change are far-reaching, including rising sea levels, extreme weather events, heatwaves, droughts, and disruptions in ecosystems and agriculture. It poses a significant threat to human livelihoods, food security, and the preservation of biodiversity. So, to avoid the destruction and the harmful effects of these societies is turning over towards renewable natural sources like hydrothermal energy, wind energy and solar energy.

The increase in the usage of human origin chemicals in soil has adverse effects on the human health and leads to the soil pollution. These pollutants consist of heavy metals and toxic organic chemicals such as pesticides, biological pathogens, and plastic waste and when these are released into water they cause hazardous effects to the aquatic life and to human beings also [10,11]. Natural materials, such as activated carbon [12], zeolites [13], and biochar [14], possess excellent adsorption properties that aid in the remediation of polluted sites. These materials can effectively remove heavy metals, organic pollutants, and other harmful substances, offering a sustainable and cost-effective approach to environmental cleanup. Natural materials like sand, gravel, and plants can be utilized to create low-tech, eco-friendly water filtration systems. Furthermore, constructed wetlands, for example, mimic natural processes to purify water and remove pollutants, making them suitable for both small-scale and large-scale applications.

Hence, due to the adverse effects caused to the environment due to the increase usage of artificial and hazardous materials, there is a need to shift toward natural materials which are cost-effective, renewable and overcome the large number of limitations offered by these materials. By embracing and advancing the use of natural materials, we can move towards a circular economy that minimizes waste, conserves resources, and promotes a harmonious coexistence with the natural world. In this chapter, we explore the diverse applications of natural materials in environmental contexts, show causing their potential to mitigate pollution, combat climate change, conserve resources, and promote sustainable living. Since, the world strives to find solutions to the pressing environmental problems of our time, the significance of natural materials cannot be overstated. By harnessing the power of nature, we can forge a path toward a greener and more resilient future.

7.2 NATURAL MATERIALS AND ENVIRONMENT

Amidst the mounting environmental challenges, the importance of natural materials has surged exponentially. Originating from renewable sources such as plants, minerals, and animals, these materials have emerged as essential elements in the pursuit of sustainable solutions. By diminishing pollution, preserving resources, and countering climate change, natural materials provide a plethora of advantages that render them invaluable for various environmental applications. One of the foremost advantages of natural materials is their renewability. Unlike finite fossil fuels, which contribute significantly to environmental degradation, natural materials are derived from resources that can be replenished over time.

This renewable characteristic ensures a more sustainable supply chain and reduces the strain on ecosystems. Ellabban et al. (2014) highlighted the global advantages of renewable energy production, categorizing them into environmental, economic, technological, social, and political aspects. Additionally, they outlined a process for the development of the renewable energy market and illustrated the obstacles that hinder the deployment of renewable energy technologies [15]. Table 7.1 shows various advantages of renewable natural sources.

Furthermore, many natural materials are biodegradable, meaning natural processes can break them down into non-toxic elements and compounds. This is in stark contrast to synthetic materials like plastics that persist for hundreds of years, causing pollution and harming wildlife [17]. Biodegradable natural materials such as bio-based plastics [18] contribute to a cleaner environment by minimizing the accumulation of non-degradable waste in landfills, oceans, and natural habitats.

The production and utilization of conventional materials often involve energy-intensive processes that contribute to greenhouse gas emissions and climate change. Moreover, to counteract this adverse trajectory, the global community is actively adopting political measures and international agreements. Notably, the recent Paris Agreement outlines a comprehensive global strategy aimed at constraining the rise in global temperatures to under 2°C [19]. Natural materials, however, generally have a lower carbon footprint throughout their life cycle. For instance, the cultivation and harvesting of plant-based materials such as bamboo or cotton require less energy and

Table 7.1
Utility of Various Renewable Energy Resources

Energy resource	Energy conversion and usage option
Hydropower	Power generation
Biomass	Heat and power generation, pyrolysis, gasification, digestion
Geothermal	Urban heating, power generation, hydrothermal, hot rock
Solar	Solar home system, solar dryers, solar cookers
Direct solar	Photovoltaic, thermal power generation, water heaters
Wind	Power generation, wind generators, windmills
Wave	Numerous designs
Tidal	Barrage, tidal stream

Source: Adapted from ABC *et al.*, copyright year publisher. Reference [16].

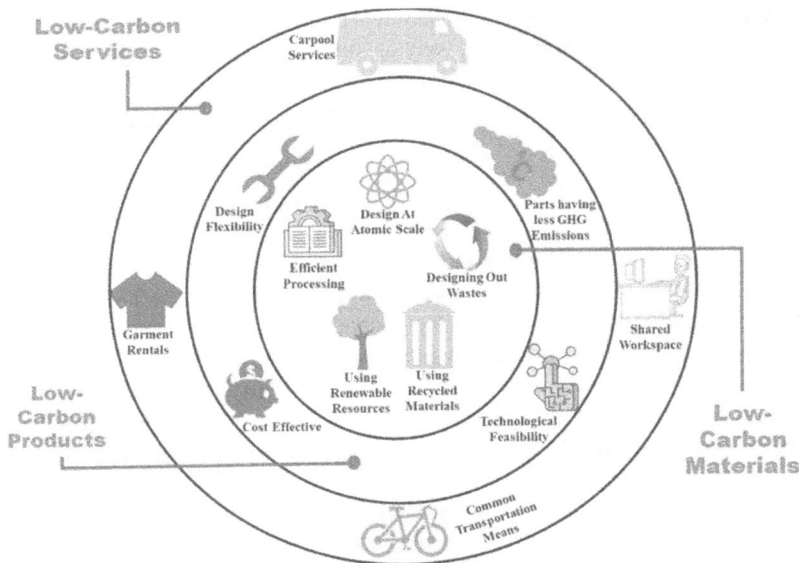

Figure 7.2 Low-carbon materials obtained with the help of low-carbon products [21]. (Adapted with permission under an open access by CC by 4.0 from Das *et al. Mater. Circ. Econ.* 3:1–11, 2021. Copyright 2021 Springer.)

emits fewer greenhouse gases compared to the extraction and processing of fossil-based materials. Natural materials can also contribute to carbon sequestration. Trees and plants, used in various applications such as construction and textiles, absorb carbon dioxide from the atmosphere during their growth, helping offset emissions and combat climate change [20]. Figure 7.2 shows various low-carbon materials obtained from low-carbon products.

Natural materials typically require fewer resources, both in terms of raw materials and energy, during extraction, processing, and manufacturing. This resource efficiency translates to less environmental impact, including lower water consumption, reduced pollution, and decreased habitat disruption. In contrast, the extraction of non-renewable resources often results in habitat destruction, soil erosion, water pollution, and the release of harmful chemicals into the environment. By using natural materials wisely, we can mitigate these negative consequences and transition toward more sustainable practices. Many synthetic materials contain harmful chemicals, such as phthalate and volatile organic compounds, which can leach into the environment and pose health risks to humans and wildlife. Natural materials, especially those that are minimally processed, are less likely to contain such toxins. This quality is particularly relevant in applications involving food storage, where the migration of harmful substances from packaging materials to food items is a concern [22, 23].

Thus, natural materials offer a holistic approach to addressing some of the most pressing environmental challenges of our time. By harnessing their renewable and biodegradable nature, minimizing carbon emissions, reducing environmental impact, and promoting healthier alternatives, we can shift toward a more sustainable future.

7.3   NATURAL MATERIALS AS BIODEGRADABLE PACKAGING MATERIALS

The proliferation of plastic packaging over the last few decades has led to an environmental crisis of colossal proportions. Plastics have revolutionized packaging due to their durability, versatility, and cost-effectiveness [24]. However, their non-biodegradable nature is a double-edged sword. The same qualities that make plastics resistant to degradation also mean that they persist in the environment for hundreds of years, causing widespread pollution and harm to ecosystems. Furthermore, it was estimated that plastic waste constitutes approximately 10% of the total municipal waste worldwide [25] and that 80% of all plastic found in the world oceans originates from land-based sources [26]. Microplastics, tiny fragments derived from the breakdown of larger plastic items, have infiltrated even the most remote areas, from the deep ocean to the highest mountains. Intentionally crafted at sizes within the millimeter or sub-millimeter range, primary microplastics are present in a wide array of everyday household items. These include personal hygiene products like facial cleansers, toothpaste, and exfoliating creams. The presence of primary microplastics in these products is particularly worrisome. Recent estimations indicate that approximately 6% of liquid skin-cleansing products available in the European Union, Switzerland, and Norway contain microplastics. Impressively, over 93% of these microplastics are comprised of polyethylene (PE) [27].

As plastics continue to accumulate in landfills, oceans, and natural habitats, there is urgency to find sustainable alternatives to protect the environment. Furthermore, biodegradable packaging materials, derived from natural sources, have emerged as a promising solution to address the plastic pollution problem while maintaining the functionalities required for packaging. These bioplastics can be biodegradable, bio-based or both as shown in Figure 7.3 [28]. Biodegradable packaging materials offer

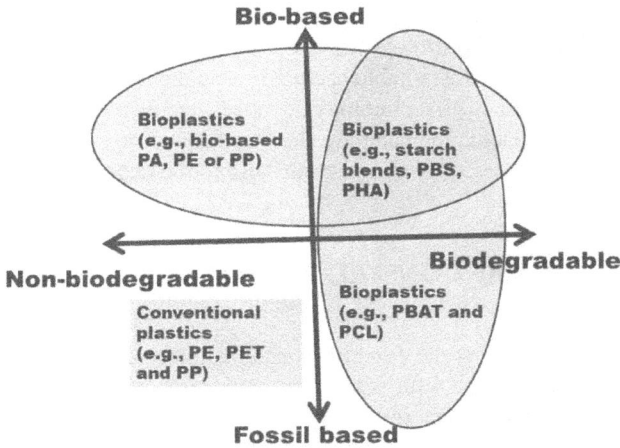

Figure 7.3 Distribution of polymers according to source materials and degree of biodegradability.

several advantages that make them attractive for environmental applications. Unlike conventional plastics, biodegradable materials break down naturally into non-toxic components over time. This significantly reduces the environmental burden caused by persistent plastic waste [29]. Biodegradable packaging materials do not accumulate as unsightly litter in the environment; whether on land or in water bodies, these materials decompose, leaving behind minimal traces. Biodegradable plastics derived from agricultural waste have the potential to serve various purposes, including packaging and salt container production, manufacturing fibers and plastic components. Moreover, there's a promising prospect of substituting non-biodegradable and petroleum-based polymers, such as polyethylene terephthalate (PET), with these environmentally friendly alternatives [30].

One of the important category of biodegradable materials is biopolymers. These are derived from renewable sources such as starch, cellulose, and proteins. These biopolymers mimic the functionality of traditional plastics while offering biodegradability. Polylactic acid (PLA), for example, is derived from corn starch and can be used to produce biodegradable cups, cutlery, and packaging films. These materials exhibit similar strength and barrier properties as conventional plastics but degrade in a much shorter time frame [31]. Another example is cellulose, which is a fundamental component of cell walls. It can be processed into transparent films that possess barrier properties against oxygen and UV light, making them suitable for packaging fresh produce [32]. Derived from the shells of crustaceans, chitosan is biodegradable, antimicrobial, and biocompatible. It can be used for various packaging applications, including antimicrobial coatings for food [33]. Starch-based materials are abundant and cost-effective. Starch can be modified to enhance its properties, resulting in materials that are water-resistant, flexible, and biodegradable [34].

In conclusion, biodegradable packaging materials hold immense promise in combating plastic pollution. From biopolymers to cellulose and starch, these materials offer a range of options for sustainable packaging that balances functionality with environmental responsibility. As technology advances and awareness grows, the adoption of biodegradable packaging can pave the way for a cleaner, more sustainable future, where packaging materials do not contribute to the degradation of our planet's ecosystems.

7.4 AS SUSTAINABLE CONSTRUCTION MATERIALS: BUILDING A GREENER FUTURE

The construction industry, while essential for human development and progress, has traditionally been associated with significant environmental impact. The extraction of non-renewable resources, energy-intensive processes, and waste generation have raised concerns about the industry's contribution to issues such as climate change and habitat destruction due to the demand for raw materials such as timber and minerals and it accounts for a significant portion of global energy consumption and greenhouse gas emission [35]. In response, sustainable construction materials have gained prominence as a means of reducing the environmental footprint of buildings and infrastructure projects. Moreover, sustainable construction materials prioritize resource efficiency, lower carbon emissions, and reduced waste generation [36].

There are various instances of sustainable construction materials which can substitute these materials. For example, bamboo being a rapidly renewable resource and can grow quickly and can be harvested without damaging the plant's root system. It has impressive strength-to-weight ratios and can be used for structural elements, flooring, and even as a concrete reinforcement [37]. Furthermore, agricultural residues are very usable in construction materials as shown in Figure 7.4. Apart from this, straw bales are used as insulation and structural elements in some sustainable construction methods. Straw is a waste product from grain crops, making it an abundant and renewable resource [38]. Certain sustainable materials, such as wood, have the added benefit of carbon sequestration. Trees absorb carbon dioxide from the atmosphere during their growth. When wood is used in construction, this carbon remains stored in the building, offsetting emissions that would otherwise be released if other materials were used.

Since these sustainable construction materials have various advantages, so many techniques are explored to make their best use such as enhancing their efficiency and impact such as prefabrication, green roofs and many more. Here, the prefabrication technique is used to reduce the construction waste and minimize the impact on the construction site [40]. On the other hand, green roofs, covered with vegetation, provide insulation, improve air quality, and manage stormwater runoff [41].

Sustainable construction materials are at the forefront of efforts to create more environmentally responsible buildings and infrastructure. By prioritizing renewable resources, reducing embodied energy, and minimizing waste, these materials play a crucial role in addressing the environmental challenges posed by the construction

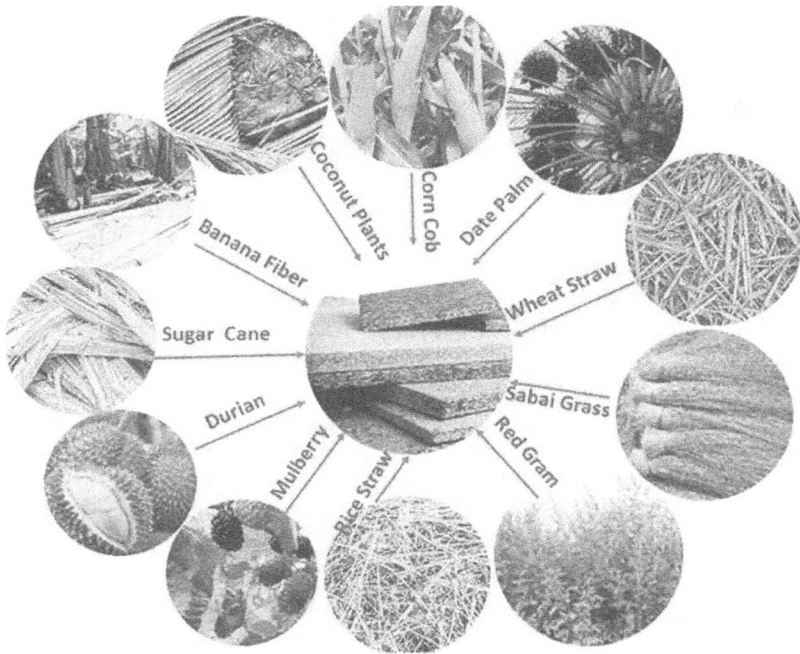

Figure 7.4 Various agricultural residues utilized in the fabrication of materials for false ceilings [39]. (Adapted with permission from Sangmesh *et al. Constr. Build, Mater.,* 368: 130457, 2023. Copyright 2023 Elsevier.)

industry. As sustainable construction practices become more widespread, they contribute not only to green buildings but also to a more sustainable and resilient built environment for the future.

7.5 REMEDIATION OF POLLUTED SITES THROUGH NATURAL PROCESSES

Pollution of soil and water resources by hazardous substances poses a significant threat to ecosystems, human health, and the overall well-being of our planet, which is caused by industrial activities, agriculture, mining, and improper waste disposal. Heavy metals, pesticides, organic pollutants, and various chemicals can accumulate in these environments, endangering aquatic life, soil fertility, and even entering the food chain, posing risks to human health [42,43]. In response to this challenge, natural remediation methods have emerged as sustainable approaches to clean up polluted sites. By harnessing the power of natural materials and processes, these methods offer effective and eco-friendly solutions for environmental cleanup.

Natural remediation methods leverage the inherent properties of certain materials to absorb, degrade, or otherwise neutralize pollutants [44]. These methods are often more sustainable and cost-effective than conventional remediation techniques that involve excavation, transport, and treatment of contaminated materials. For instance, activated carbon which is a highly porous material with a large surface area, making it an effective adsorbent for a wide range of contaminants. It can adsorb organic compounds, heavy metals, and even some chemicals that contribute to groundwater contamination. Activated carbon is used in various forms, such as granules or powders, in treatment systems to remove pollutants from both air and water [45]. Another example being zeolites which can be considered as a synthetic as well as natural materials and have a unique porous structure that enables them to trap and exchange ions. They are used in environmental remediation to adsorb heavy metals, ammonia, and other pollutants. Zeolites can be incorporated into permeable barriers to capture contaminants in groundwater, or they can be used in filters to treat water and wastewater [46] as shown in Figure 7.5. The natural zeolites are capable of interacting cationic species while to remove organic and anionic species, these zeolites are modified [47]. Biochar, a carbon-rich material produced from the pyrolysis of organic matter, has gained attention for its potential in soil remediation. Biochar can improve soil structure, retain nutrients, and immobilize contaminants. Its porous structure provides a habitat for beneficial microorganisms that can aid in the degradation of pollutants over time especially arsenic and cadmium from the soil [48].

Natural remediation methods offer a sustainable and efficient approach to cleaning up polluted sites. By harnessing the capabilities of natural materials, these methods mitigate the harmful effects of pollutants on the environment and human health. As

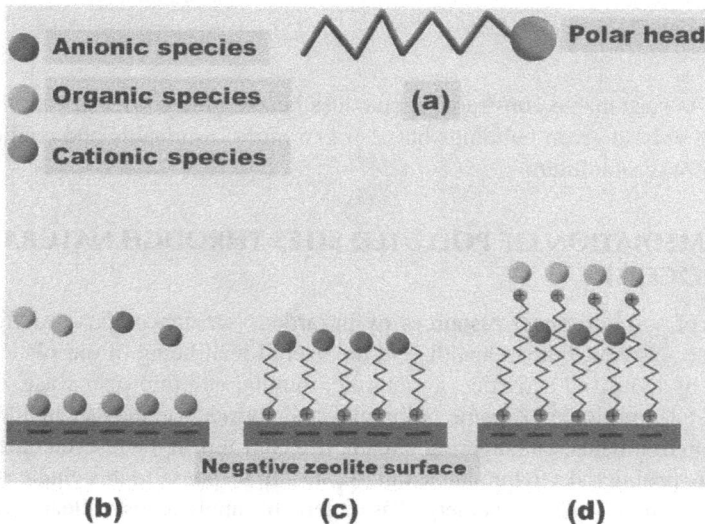

Figure 7.5 (a) Polar head. Removal of (b) cationic, (c) anionic, and (d) organic species by zeolites [47]. (Adapted with permission under an open access by CC by 4.0 from de *et al.*, *Adsorpt. Sci. Technol.*, 1–26, 2022. Copyright 2022 Hindawi Limited.)

we continue to explore and refine these techniques, natural remediation holds the potential to transform contaminated areas into healthier, more sustainable ecosystems.

7.6 BIO-BASED FUELS AND ENERGY SOURCES

The burning of fossil fuels, such as coal, oil, and natural gas, releases greenhouse gases into the atmosphere, contributing significantly to global warming and climate change. To curb these environmental impacts, the world is shifting toward renewable energy sources that generate minimal or no greenhouse gas emissions during their production and use. Bio-based fuels and energy sources harness the power of nature to provide cleaner, renewable options for meeting our growing energy demands. Biofuels are fuels derived from biological sources, such as plants and algae, collectively known as biomass. Biomass is a renewable resource that can be sustainably harvested and cultivated. It has become a focal point for bio-based energy solutions due to its potential to replace traditional fossil fuels [49]. There are three generations of biofuels as shown in Table 7.2.

There are various sources for the production of biofuels, for instance, algae holds immense potential as a bio-based energy source due to their rapid growth and high oil content. Algae can be cultivated in various environments, including ponds, tanks, and even wastewater treatment facilities [51]. Algae biomass can be converted into biofuels such as biodiesel and bioethanol, as well as valuable co-products like animal feed and bioplastics [52]. The procedure of conversion of algae into biofuels is depicted in Figure 7.6. Agricultural and forestry residues, along with organic waste, represent vast sources of biomass that can be converted into energy. Processes such as anaerobic digestion and fermentation can transform these materials into biogas,

Table 7.2
Classification of Biofuels Based on Various Generations

Generation	Description	Example
First	Biofuels from traditional energy crops (sugar, starch, and oilseed) and animal fat	*Ethanol*: Corn, wheat, and rice *Biodiesel*: Rapeseed, sunflower, and soyabean
Second	Biofuels from non traditional energy crops (sugar, starch, and oilseed); from perennial grasses, agricultural and forestry wastes and lignocellulosic species	*Ethanol*: Switchgrass (*Panicum virgatum*) and sweet sorghum *Biodiesel*: Jatropha and castor bean
Third	Biofuels from micro and macro algae It is used most advanced technology for processing and production	*Ethanol*: *Chlamydomonas reinhardti* and *Chlorococcum littorale* *Biodiesel*: *Oedogonium* sp. and *Spirogyra* sp.

Source: Reference [50].

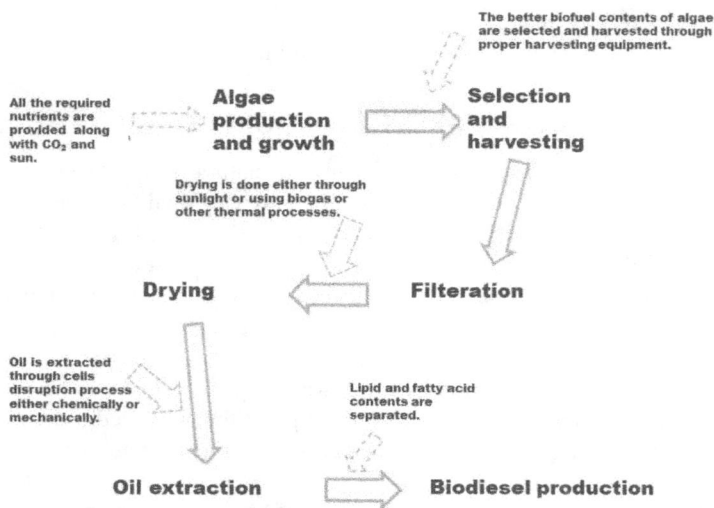

The better biofuel contents of algae
are selected and harvested through
proper harvesting equipment.

All the required
nutrients are
provided along
with CO₂ and
sun.

**Algae
production
and growth**

**Selection
and
harvesting**

Drying is done either through
sunlight or using biogas or
other thermal processes.

Drying   **Filteration**

Oil is extracted
through cells
disruption process
either chemically or
mechanically.

Lipid and fatty acid
contents are
separated.

Oil extraction   **Biodiesel production**

Figure 7.6 Process of conversion of algae into biofuels [53]. (Adapted with permission under an open access by CC BY-NC-ND 4.0 from Khan *et al. Hayati. J. Biosci.* 24:163–167, 2017. Copyright 2017 Elsevier.)

which primarily consist of methane and carbon dioxide. Biogas can be used for heating, electricity generation, and even as a vehicle fuel.

While bio-based fuels hold promise for reducing greenhouse gas emissions, their environmental impact is complex and context-dependent. Factors such as land use change, water consumption, and the energy required for cultivation and processing can influence their overall sustainability. Bio-based fuels and energy sources play a crucial role in the transition to a sustainable energy future. By reducing greenhouse gas emissions, promoting resource efficiency, and offering alternatives to finite fossil fuels, these solutions contribute to mitigating climate change and ensuring energy security [54]. However, it's essential to approach their development with a holistic perspective, considering not only their benefits but also potential trade-offs and long-term impacts on ecosystems, food systems, and land use. In conclusion, bio-based fuels and energy sources represent a vital pillar in the global effort to address climate change and establish a more sustainable energy landscape. By tapping into the abundance of nature's resources, we can forge a pathway toward cleaner, renewable energy that supports the needs of present and future generations while safeguarding the planet.

7.7 NATURAL WATER FILTRATION SYSTEMS

Rapid urbanization, industrialization, and agricultural activities have led to contamination of water sources with pollutants such as heavy metals, pathogens, and chemicals. Traditional water treatment methods can be resource-intensive and often fall short in providing safe water to all communities. Natural water filtration systems offer an alternative approach rooted in ecological principles. Natural water filtration systems utilize a combination of physical, chemical, and biological processes to remove contaminants from water. These systems often rely on locally available and renewable materials to achieve effective filtration. One of the most effective method is to use sand and gravel as filters [55]. Sand and gravel have natural filtration properties due to their particle size distribution and pore spaces. These materials trap and remove suspended solids, pathogens, and even some dissolved contaminants as water percolates through the filter bed.

Certain plants, known as hyperaccumulators, have the ability to absorb and accumulate heavy metals and pollutants from water and soil. These plants can be used in constructed wetlands and natural water bodies to remove contaminants [56]. Plant roots provide a habitat for beneficial microorganisms that contribute to the breakdown of organic matter and the removal of contaminants. The movement of water through the plant roots introduces oxygen, promoting aerobic conditions that support biological activity. Furthermore, some plants absorb excess nutrients, such as nitrogen and phosphorus, from water, reducing the risk of eutrophication in receiving water bodies.

7.7.1 APPLICATIONS OF NATURAL WATER FILTRATION SYSTEM

Natural water filtration systems have gained traction as eco-friendly alternatives to conventional water treatment methods. These systems utilize natural processes, materials, and ecosystems to purify water, offering a range of applications that address diverse water quality challenges. From small-scale solutions in rural areas to large-scale urban water management, natural water filtration has demonstrated its feasibility and effectiveness in various contexts.

7.7.1.1 Small-Scale Applications

Natural water filtration systems can be implemented at the household level, especially in areas without access to centralized water treatment facilities. Simple sand and gravel filters or biochar-based filtration systems can be used to remove sediments, pathogens, and impurities from drinking water. Constructed wetlands or vegetated filter strips can be integrated into community water systems to treat both drinking water and waste water. These systems provide a decentralized approach to water treatment, reducing the strain on centralized infrastructure [57].

7.7.1.2 Urban Water Management

Urban areas generate significant stormwater runoff, which often carries pollutants from roads, roofs, and other surfaces. Constructed wetlands, bio-infiltration basins, and vegetated swales can help capture, filter, and purify storm water before it enters natural water bodies. Grey water, the relatively clean waste water generated from activities like bathing and laundry, can be treated and recycled for non-potable uses such as landscape irrigation and toilet flushing. Natural systems like sand filters and constructed wetlands can effectively purify grey water for reuse [58].

7.8   THE POWER OF NATURAL MATERIALS IN TEXTILE INDUSTRY

The textile industry, long celebrated for its creativity and innovation, is now facing a crucial imperative sustainability. The unsustainable practices that have historically characterized the industry, from excessive water usage to pollution and labor exploitation, are being reevaluated in the wake of a global call for environmental and social responsibility. These textiles, woven with a commitment to ecological harmony and ethical practices, are catalyzing a paradigm shift toward a greener and more equitable fashion landscape. Many materials like organic cotton, hemp are being effectively used to ensure environment protection. Since organic cotton is grown without synthetic pesticides and fertilizers, it reduces environmental impact and minimizes harm to farmers and ecosystems [59]. Moreover, unlike the conventional cotton, organic cotton remains gentle on delicate infant skin. It stands as an ideal choice for swaddling, cleansing, and caring for newborns. Its suitability extends to crafting baby clothing, bandages, wound care, crib bedding, adorable outfits, soft towels, and an array of essential items for babies. Furthermore, hemp cultivation requires minimal water and pesticides, making it a sustainable alternative. Hemp fibers are strong, durable, and biodegradable [60]. Moving further, natural dyes obtained from plants, such as indigo, turmeric, and madder root, are being used to replace synthetic dyes that often involve harmful chemicals. These natural dyes not only offer vibrant colors but also reduce water pollution and health hazards [61]. The emergence of biofabricated textiles, where materials are grown from living organisms, presents a novel way of producing sustainable textiles. This includes materials like mycelium (mushroom-based) leather and lab-grown fibers.Sustainable textiles embody a profound transformation in the fashion industry. They signify a departure from the take-make-dispose approach towards a circular and regenerative model. By embracing sustainable textiles, we embark on a journey that transcends fashion and becomes a powerful agent of positive change, empowering us to make choices that resonate with the values of a responsible and compassionate world.

7.9   CONCLUSION AND FUTURE PROSPECTS

The realm of environmental applications has found a true ally in the form of natural materials, forging a path towards a sustainable and harmonious coexistence between human activities and the planet's intricate ecosystems. The exploration of natural water filtration systems underscores the immense potential for harnessing nature's mechanisms to address pressing water quality challenges. These systems offer a myriad of benefits, from providing access to clean water and mitigating pollution to restore ecosystems and promoting environmental sustainability. Natural water filtration systems have demonstrated their effectiveness across various scales and contexts. They have showcased their potential to provide access to clean water, reduce environmental impact of hazardous pollutants and surging the sustainability in ecosystem. Natural materials, with their inherent biodegradability, minimal ecological footprint, and diverse functionalities, have heralded a new era of conscious consumption and responsible production. From biodegradable packaging materials that combat plastic pollution to sustainable construction materials that reduce the carbon footprint of buildings, these applications demonstrate the convergence of technological innovation and environmental stewardship. The multifaceted benefits they offer, including reduced waste, lower resource consumption, and enhanced ecosystem health, underscore their pivotal role in the pursuit of sustainability. As we gaze towards the future, the prospects of environmental applications of natural materials are ripe with potential. The ongoing development of novel biodegradable materials, the refinement of natural remediation techniques, and the integration of these solutions into mainstream practices hold the promise of an even more sustainable world. Research, collaboration, and continued innovation will propel us towards a future where natural materials stand as the cornerstone of a circular economy, where waste is minimized, resources are optimized, and ecosystems flourish. By harnessing the innate qualities of nature's offerings, we embark on a journey of transformation that has the power to reshape industries, societies, and ultimately, the world we leave for generations to come.

REFERENCES

1. L. Markovic, A. Celinscak, Z. Vlaovic, I. Grbac, and D. Domljan. Importance of natural materials for good and quality sleep. 12 2019.
2. S. Burroughs and J. Ruzicka. The use of natural materials for construction projects social aspects of sustainable building: Case studies from australia and europe. *IOP Conference Series: Earth and Environmental Science*, **290**:012009, 06 2019.
3. I. Manisalidis, E. Stavropoulou, A. Stavropoulos, and E. Bezirtzoglou. Environmental and health impacts of air pollution: A review. *Frontiers in Public Health,* **8**:14, 2020.
4. A. O. C. Iroegbu, S. Sinha Ray, V. Mbarane, J. C. Bordado, and J. Paulo Sardinha. Plastic pollution: A perspective on matters arising: Challenges and opportunities. *ACS Omega*, **6**(30):19343–19355, 2021.
5. C. G. Alimba and C. Faggio. Microplastics in the marine environment: Current trends in environmental pollution and mechanisms of toxicological profile. *Environmental Toxicology and Pharmacology*, **68**:61–74, 2019.
6. Q. Zhou, H. Zhang, C. Fu, Y. Zhou, Z. Dai, Y. Li, C. Tu, and Y. Luo. The distribution and morphology of microplastics in coastal soils adjacent to the bohai sea and the yellow sea. *Geoderma*, **322**:201–208, 2018.
7. M. Sreekala and S. Thomas. Effect of fibre surface modification on water-sorption characteristics of oil palm fibres. *Composites Science and Technology*, **63**(6):861–869, 2003.
8. S. S. Hiremath et al. Natural fiber reinforced composites in the context of biodegradability: A review. 2020.
9. S. Izzah, C. Blessya, and E. Iriyo. The potential of cocoa bean shell waste for bioethanol to support energy transition in cocoa plantation central sulawesi. 08 2021.
10. K. E. Cosselman, A. Navas-Acien, and J. D. Kaufman. Environmental factors in cardiovascular disease. *Nature Reviews Cardiology*, **12**(11):627–642, 2015.
11. T. Münzel, T. Gori, S. Al-Kindi, J. Deanfield, J. Lelieveld, A. Daiber, and S. Rajagopalan. Effects of gaseous and solid constituents of air pollution on endothelial function. *European Heart Journal*, **39**(38):3543–3550, 2018.

12. M. S. Shafeeyan, W. M. A. W. Daud, A. Houshmand, and A. Shamiri. A review on surface modification of activated carbon for carbon dioxide adsorption. *Journal of Analytical and Applied Pyrolysis*, **89**(2):143–151, 2010.

13. S. M. Shaheen, A. S. Derbalah, F. S. Moghanm, et al. Removal of heavy metals from aqueous solution by zeolite in competitive sorption system. *International Journal of Environmental Science and Development*, 3(4):362–367, 2012.

14. X. Tong and R. Xu. Removal of Cu(II) from acidic electroplating effluent by biochars generated from crop straws. *Journal of Environmental Sciences*, 25(4):652–658, 2013.

15. O. Ellabban, H. Abu-Rub, and F. Blaabjerg. Renewable energy resources: Current status, future prospects and their enabling technology. *Renewable and Sustainable Energy Reviews*, **39**:748–764, 2014.

16. A. Demirbaş. Global renewable energy resources. *Energy sources*, **28**(8):779–792, 2006.

17. H. Saygin and A. Baysal. Similarities and discrepancies between bio-based and conventional submicron-sized plastics: in relation to clinically important bacteria. *Bulletin of Environmental Contamination and Toxicology*, **105**:26–35, 2020.

18. Z. S. Mazhandu, E. Muzenda, T. A. Mamvura, M. Belaid, and T. Nhubu. Integrated and consolidated review of plastic waste management and bio-based biodegradable plastics: Challenges and opportunities. *Sustainability*, **12**(20):8360, 2020.

19. Commissione Europea. Accordo di parigi. *Energia, cambiamenti climatici, ambiente, link: https://ec. europa. eu/clima/policies/international/negotiations/paris 'it*, 2016.

20. A. Favier, C. D. Wolf, K. Scrivener, and G. Habert. A sustainable future for the european cement and concrete industry: Technology assessment for full decarbonisation of the industry by 2050. Technical report, ETH Zurich, 2018.

21. O. Das, Á. Restás, V. Shanmugam, G. Sas, M. Försth, Q. Xu, L. Jiang, M. S. Hedenqvist, and S. Ramakrishna. Demystifying low-carbon materials. *Materials Circular Economy*, **3**:1–11, 2021.

22. N. Ahmed, J. Singh, H. Kour, and P. Gupta. Naturally occurring preservatives in food and their role in food preservation. *International Journal of Pharmaceutical and Biological Archive*, **4**:22–30, 2013.

23. O. A. Fatoki and Deborah A. Onifade. Use of plant antimicrobials for food preservation. *International Journal of Bioengineering and Life Sciences*, **7**(12):1110–1113, 2013.

24. J. P. d. Costa, P. S. Santos, A. C. Duarte, and T. Rocha-Santos. (Nano) plastics in the environment—sources, fates and effects. *Science of the Total Environment*, **566**:15–26, 2016.

25. J. R. Jambeck, R. Geyer, C. Wilcox, T. R. Siegler, M. Perryman, A. Andrady, R. Narayan, and K. L. Law. Plastic waste inputs from land into the ocean. *Science*, **347**(6223):768–771, 2015.

26. J. Da Costa, A. Duarte, and T. Rocha-Santos. *Microplastics Occurrence, fate and behaviour in the Environment*. 2016.

27. T. Gouin, J. Avalos, I. Brunning, K. Brzuska, J. D. Graaf, J. Kaumanns, T. Koning, M. Meyberg, K. Rettinger, H. Schlatter, et al. Use of micro-plastic beads in cosmetic products in Europe and their estimated emissions to the north sea environment. *SOFW J*, **141**(4):40–46, 2015.

28. Ana Paço, Jéssica Jacinto, João Pinto da Costa, Patrícia S.M. Santos, Rui Vitorino, Armando C. Duarte, and Teresa Rocha-Santos. biotechnological tools for the effective management of plastics in the environment. *Critical Reviews in Environmental Science and Technology*, **49**(5):410–441, 2019.

29. J. Zhu and C. Wang. Biodegradable plastics: Green hope or greenwashing? *Marine Pollution Bulletin*, **161**:111774, 2020.
30. N. Mostafa, A. A. Farag, H. M. Abo-dief, and A. M. Tayeb. Production of biodegradable plastic from agricultural wastes. *Arabian journal of chemistry*, **11**(4):546–553, 2018.
31. J. d. M. Fonseca?, B. L. Koop?, T. C. Trevisol, C. Capello, A. R. Monteiro, and G. A. Valencia. An overview of biopolymers in food packaging systems. *Nanotechnology-Enhanced Food Packaging*, pages 19–53, 2022.
32. H. Sadeghifar, R. Venditti, J. Jur, R. E. Gorga, and J. J. Pawlak. Celluloselignin biodegradable and flexible uv protection film. *ACS Sustainable Chemistry & Engineering*, **5**(1):625–631, 2017.
33. M. Flórez, E. Guerra-Rodríguez, P. Cazón, and M. Vázquez. Chitosan for food packaging: Recent advances in active and intelligent films. *Food Hydrocolloids*, **124**:107328, 2022.
34. R. Frische, R. Gross-Lannert, K. Wollmann, B. Best, E. Schmid, and F. Buehler. Water-resistant starch materials for the production of cast sheets and thermoplastic materials. *Journal of Cleaner Production*, **2**(4):127, 1996.
35. S. O. Ametepey and S. Kwame Ansah. Impacts of construction activities on the environment: the case of ghana. *Journal of Construction Project Management and Innovation*, **4**(sup-1):934–948, 2014.
36. P. Dräger and P. Letmathe. Value losses and environmental impacts in the construction industry–tradeoffs or correlates? *Journal of Cleaner Production*, **336**:130435, 2022.
37. M. Guan, C. Yong, L. Wang, and Q. Zhang. Selected properties of bamboo scrimber flooring made of India Melocanna Baccifera. In *Proceedings of the 55th International Convention of Society of Wood Science and Technology, Beijing, China*, 2012.
38. M. Pierzchalski. Straw bale building as a low-tech solution: A case study in northern poland. *Sustainability*, **14**(24):16511, 2022.
39. B. Sangmesh, N. Patil, K. K. Jaiswal, T. P. Gowrishankar, K. Karthik Selvakumar, M. Jyothi, R. Jyothilakshmi, and S. Kumar. Development of sustainable alternative materials for the construction of green buildings using agricultural residues: A review. *Construction and Building Materials*, **368**:130457, 2023.
40. D. J. Tony and R. Kokila. Study on prefabrication technique in construction and its barriers. *The Asian Review of Civil Engineering*, **7**(1), 2018.
41. R. M. Ahmed and H. Z. Alibaba. An evaluation of green roofing in buildings. *International Journal of Scientific and Research Publications*, **6**(1):366–373, 2016.
42. A. Cuypers, M. Plusquin, T. Remans, M. Jozefczak, E. Keunen, H. Gielen, K. Opdenakker, A. Ravindran Nair, E. Munters, T. J. Artois, et al. Cadmium stress: An oxidative challenge. *Biometals*, **23**:927–940, 2010.
43. I. R. Lake, L. Hooper, A. Abdelhamid, Graham Bentham, Alistair B. A. Boxall, Alizon Draper, Susan Fairweather-Tait, Mike Hulme, Paul R Hunter, Gordon Nichols, et al. Climate change and food security: health impacts in developed countries. *Environmental Health Perspectives*, **120**(11):1520–1526, 2012.
44. Alfred M Beeton. Changes in the environment and biota of the great lakes. 1969.
45. M. Marton, J. Ilavskỳ, and D. Barloková. Adsorption of specific chloroacetanilides on granular activated carbon. In *IOP Conference Series: Materials Science and Engineering*, volume 867, page 012031. IOP Publishing, 2020.
46. M. R. Adam, M. H. D. Othman, S. K. Hubadillah, M. H. Abd Aziz, and M. R. Jamalludin. Application of natural zeolite clinoptilolite for the removal of ammonia in wastewater. *Materials Today: Proceedings*, 2023.

47. L. F. d. Magalhães, G. R. d. Silva, and A. E. C. Peres. Zeolite application in wastewater treatment. *Adsorption Science & Technology*, **2022**:1–26, 2022.
48. M. Qiu, L. Liu, Q. Ling, Y. Cai, S. Yu, S. Wang, D. Fu, B. Hu, and X. Wang. Biochar for the removal of contaminants from soil and water: A review. *Biochar*, **4**(1):19, 2022.
49. M. A. D. S. Bernardes. Biofuel production: Recent developments and prospects. 2011.
50. C. Román-Figueroa and M. Paneque. Ethics and biofuel production in Chile. *Journal of Agricultural and Environmental Ethics*, **28**:293–312, 2015.
51. G. W. Roberts, M. P. Fortier, B. S. Sturm, and S. M. Stagg-Williams. Promising pathway for algal biofuels through wastewater cultivation and hydrothermal conversion. *Energy & Fuels*, **27**(2):857–867, 2013.
52. V. Patil, K. Tran, and H. R. Giselrød. Towards sustainable production of biofuels from microalgae. *International journal of molecular sciences*, **9**(7):1188–1195, 2008.
53. S. Khan, R. Siddique, W. Sajjad, G. Nabi, K. M. Hayat, P. Duan, and L. Yao. Biodiesel production from algae to overcome the energy crisis. *HAYATI Journal of Biosciences*, **24**(4):163–167, 2017.
54. F. A. Malla and S. A. Bandh. Biofuels and sustainable development goals. In *Environmental Sustainability of Biofuels*, pages 13–26. Elsevier, 2023.
55. R. Mohamed, M. N. Adnan, M. A. Mohamed, and A. M. Kassim. Conventional water filter (sand and gravel) for ablution water treatment, reuse potential, and its water savings. *Journal of Sustainable Development*, **9**(1):35–43, 2016.
56. P. Papadia, F. Barozzi, D. Migoni, M. Rojas, F. P. Fanizzi, and G.-P. D. Sansebastiano. Aquatic mosses as adaptable bio-filters for heavy metal removal from contaminated water. *International Journal of Molecular Sciences*, **21**(13):4769, 2020.
57. M. O. Mohammed and A. A. Solumon. Two models of household sand filters for small scale water purification. *Polish Journal of Environmental Studies*, **31**:2737–2748, 2022.
58. A. Katukiza, M. Ronteltap, C. Niwagaba, F. Kansiime, and P. Lens. Grey water treatment in urban slums by a filtration system: Optimisation of the filtration medium. *Journal of Environmental Management*, **146**:131–141, 2014.
59. R. A. Angelova. Organic cotton: Technological and environmental aspects. In Proc. of XXIII Int. Scientific Conference FPEPM, September, pages 17–20, 2018.
60. L. Kramer. Hemp as a raw material for the fashion industry a stud on determining major factors hampering hemp to be integrated in the textile apparel supply chain. *Research Gate*, 2017.
61. H. Křížová. Natural dyes: Their past, present, future and sustainability. Recent Developments in Fibrous Material Science. Prague: Kosmas Publishing, pages 59–71, 2015.

8 X-ray Absorption Spectroscopy of Calcium-Based Natural Materials

Sanjeev Gautam
Advanced Functional Materials Laboratory, Dr. S.S. Bhatnagar
University Institute of Chemical Engineering and Technology
Panjab University, Chandigarh, India

Monika Verma and Vishal Thakur
Advanced Functional Materials Laboratory, Dr. S.S. Bhatnagar
University Institute of Chemical Engineering and Technology and
Energy Research Centre, Panjab University, Chandigarh, India

Mandeep Kaur
Advanced Functional Materials Laboratory, Dr. S.S. Bhatnagar
University Institute of Chemical Engineering and Technology and
Department of Physics, Panjab University, Chandigarh, India

Ramjanay Chaudhary and Mukul Gupta
UGC DAE Consortium for Scientific Research, Indore, India

8.1 X-RAY ABSORPTION SPECTROSCOPY

Calcium found in abundance in nature in different forms like calcium carbonate ($CaCO_3$), calcium phosphate ($Ca_3(PO_4)_2$), calcium sulphate ($CaSO_4$), calcium oxide (CaO), calcium silicate ($CaSiO_3$), calcium aluminates, calcium hydroxide, etc. These are some examples only of calcium based natural materials, also it has wide properties in industries, in water treatment plants, in biological field. One of the examples is calcium orthophosphates, such as hydroxyapatite ($Ca_5(PO_4)_3OH$) and monetite ($CaHPO_4$), are widely used as a source of phosphorus which is used in agricultural fertilizers, as detergents and cleaning agents phosphorus chemicals, in dental applications. Calcium silicate is a natural obtained materials used in cement factory and the synthesis of nano silicate [1] that can be used as an additive to increase the durability and mechanical property of concrete. To study the structure of calcium based natural materials for developing and optimizing various application, XAS spectroscopy [2–4] gives insight of structure, coordination environment, oxidation state, bond length, bond angle, etc. which prove to be a powerful technique because of its sensitivity to local electronic and atomic structure, as well as its element-specific nature. Its applications can be seen in various fields, including catalysis, materials science, environmental science, and biochemistry. Interpretation of XAS spectra typically involves comparing experimental data with theoretical calculations, such as density functional theory (DFT) calculations, to obtain quantitative information about the sample's atomic and electronic structure.

8.1.1 BASIC PRINCIPLE AND WORKING

A technique uses to study the electronic structure of atoms and molecules consists of heavy metal for producing X-rays, monochromator, collimator, and sample holders. When a photon of certain energy hv known as photon incident on the compound an electron in the core level is excited into unoccupied atomic/molecular orbitals above the fermi level. It provides valuable information about the local atomic environment, oxidation state, and coordination geometry of elements in a sample. Different modes like total electron yield (TEY) mode, transmission mode [5], fluorescence yield (FEY) mode [6], and partial electron yield (PEY) mode [7], are well known for XAS spectrum's record. Transmission mode in which X-ray beam's intensity just before and after the transmission is measured. It is not suitable for thin film as X-ray can only travel to certain depth of the substances and if thin film and substrate both are of same substances than the transmitted ray is hard to differentiate. But it is well known for hard X-ray where the mean absorption of lengths is long. For thin films, other mode like PEY, TEY, and FEY can be used for the same, although it depends on the nature of materials and the type of materials for which XAS can be used.

Two main parts of Figure 8.1, where

1. X-ray Absorption Near-Edge Structure (XANES)
2. Extended X-ray Absorption Fine Structure (EXAFS)

where XANES spectroscopy focuses on the absorption region near the absorption edge of an element that gives information about the oxidation state and coordination environment of the absorbing atom. Multiple scattering processes resulted from the low kinetic energy of photoelectron in the region enhanced XANES absorption spectrum that improves the structural determination by comprehending the bond angle information. XANES is particularly useful for identifying chemical species and determining the local electronic structure of a sample. EXAFS spectroscopy examines the fine structure of the X-ray absorption spectrum beyond the absorption edge and tells us about the local atomic arrangement around the absorbing atom, including bond lengths, coordination numbers, and bond angles. EXAFS is often employed to study the atomic structure and dynamics of materials, such as catalysts, nanoparticles, and complex compounds [8].

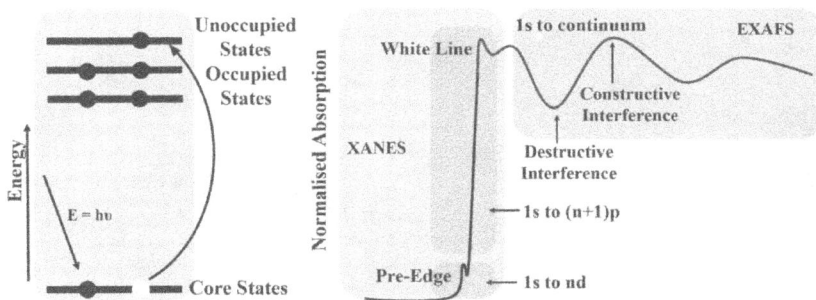

Figure 8.1 Phenomena of XAS and its two category of absorption XANES and EXAFS.

8.1.2 CA K-EDGE ABSORPTION

Abundance of natural occur materials containing the calcium possesses different biological and industrial applications. Like natural occur garnet have complex solution $X_3Y_2Z_3O_{12}$, where X, Y, and Z represent different materials X = Mg, Fe, Ca, Mn; Y = Al, Fe, Cr; and Z = Si, respectively [9]. XRD of garnet single crystal showed the presence of compositional gap in the range of 1.4–1.2 Ca atoms per formula unit (p.f.u.), but simple XRD can not tell anything about local environment of a specific chemical species. Thus using XANES, in order to get information about local environment.

XANES spectra at the Ca-K edge for various Ca occupancy at the X site [10] of complex solution of garnet, with 100% occupancy corresponding to 3.0 Ca (p.f.u.). Approximately 50 eV energy range above the edge, XANES spectra exhibit up to nine distinct structures. Among these structures, seven occur at first 30 eV from A to G with the increment in energy can be seen in Figure 8.2, also the energy position and intensity of these features vary across the series of compounds, depending on concentration of specific Ca. In experimental spectra, at pre-edge for A feature energy position and intensity seems to be independent of Ca content in garnet. Most intense spectra observed with energy increased resulting from $1s$-np transition where shoulder peak (B) remains unchanged and two peak (C and D) where relative intensity may vary due to substitution of Fe at X site. Another peak (E) which can be disappeared at low concentration of Ca content.

Ratio of $\frac{C}{D}$ intensity peak gives information about the local environment of Ca that belongs to grossular and almandine series; however, it could change not frequently in series depending upon the substitution of Ca-Fe. All the calculations of XANES spectra using on the multiple-scattering CONTINUUM code based on single electron full multiple scattering [11–13].

Ca present in soil present in carbonate, phosphate [14,15], silicate halogenide and in organic form, compounds present in soil shows largest peak (white line) at 4050 $\pm$ 1 eV except calcite and dolomite that shows at 4048.5 and 4048.3 eV, respectively. All compounds pre-edge peak/shoulder appear at $\sim$ 4040–4041 eV [16].

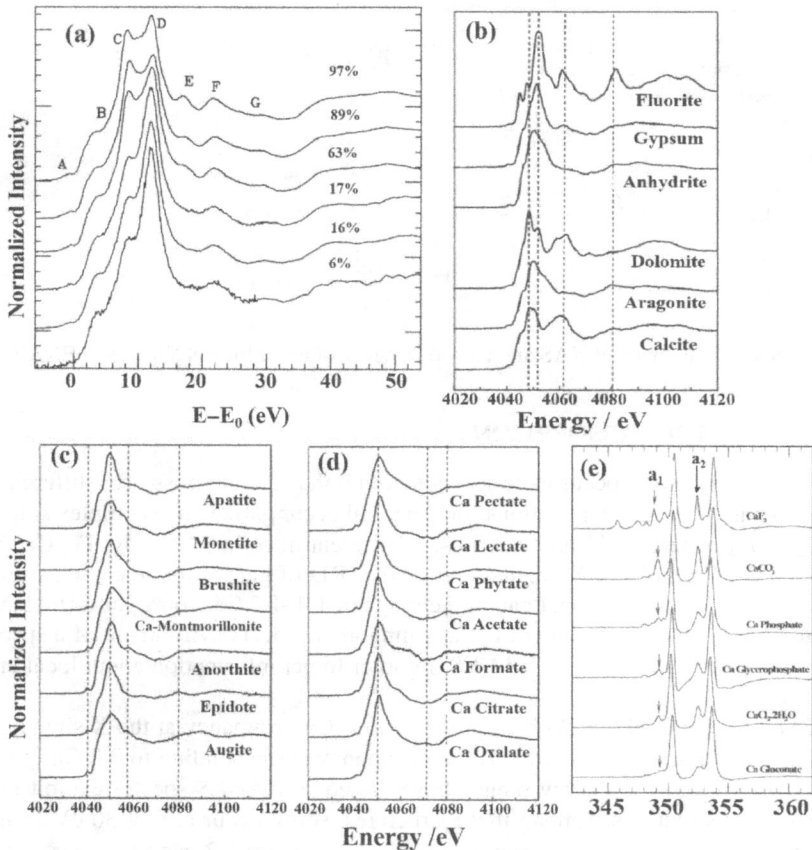

Figure 8.2 Experimental XANES spectra of Ca-K edge natural materials, (a) Ca-content garnets [13], (Adapted with permission from J. Chaboy *et al. Phys. Rev. B*, 52:6349, 1995. Copyright 1995 APS.) (b) Shows the compound carbonate, sulfates, and fluorite where Ca has been extracted. (c) Shows Ca from silicates and phosphate sources found in soil, (d) Shows organic compound exist in soil and are bounded to Ca [16]. (Adapted with permission from J. Prietzel *et al. Biochemistry*, 152:195–222, 2021. Copyright 2021 Springer.) (e) Shows the experimental L-edge spectra of XANES Ca compound [17]. (Adapted from S. J. Naftel *et al. J. Synchrotron Radiat.* 8:255–257, 2001. Copyright 2001 International Union of Crystallography.)

8.1.3 CA L-EDGE ABSORPTION

Ca L-edge absorption of some compound has been shown in Figure 8.2(e), where peaks are arising due to the symmetry of atoms surrounding the Ca^{+2} in the first co-ordination sphere. Peaks resulting from energy separation of splitting of $2p$ spin-orbit and $2d$ with the crystal field produces $L_{3,2}$ XANES peaks. Various compound like CaF_2 [18], $CaCO_3$, Ca phosphate, Ca Glycerophosphate, $CaCl_2 \cdot H_2O$,

Ca Gluconate shows the largest peak at $\sim$350–355 eV and these peaks depend on the environment and the symmetry of Ca^{+2} ion. Among the compounds, $CaCl_2 \cdot H_2O$ show coordination number 6, $CaCO_3$ and CaF_2 shows coordination number 6 and 8, respectively. Ca Glycerophosphate and Ca phosphate show almost identical spectra suggesting the same structure of Ca site. Intensity of spectra provides vital information regarding magnitude of crystal field as from relative intensity of (a_1) and (a_2) of Ca Gluconate and CaF_2 have smallest and largest crystal field parameter [17]. Another Ca compounds shows the XANES of calcite, $CaCO_3$, dolomite, aragonite where the spectrum is identical to calcite and again show the environment of Ca^{+2} in a compound [19].

8.2 X-RAY PHOTOELECTRON SPECTROSCOPY

The knowledge and understanding of the surface chemical constitution, interface, and structure is very significant for knowing the corrosion properties of different materials in various conditions of environments. A few nanometer thickness of the layer surface take part in rapid surface reaction with the atmosphere, hence the instruments capable of examine surface layers upto nanoscale are required. XPS is the favorable tool to probe chemical proportion of material's surface (within 10 nm), also the information of elements and its detection sensitivity of range 0.1–1 at.% for every periodic elements excluding H and He. Hence, these properties make XPS a special tool for corrosion evaluation and surface characterization [20,21]. It is a powerful instrument for the study of surface reactions (vacuum) upto a monolayer. The spectra of XPS are derived by illuminating a material with a X-ray beam (like Mg K_α or Al α) along with finding of kinetic energy (K.E.) value and the total number of electrons escaped from atoms on the surface of material being examined [22]. The spectra of XPS is obtained by utilizing electron intensity that escapes from the material's surface and value of binding energy (B.E.) (calculation of B.E. can be done from the recorded kinetic energy). The ultra-high vacuum (UHV) system is required in XPS surface analysis to acquire the maximum count of electron during spectra collection. Usually, the analyzer is kept 1 m away distance from the surface of X-ray illumination. This UHV surface analytical approach evaluates the quantitative and qualitative elemental composition having an atomic number of lithium element and above elements as well as used for the identification of the type of chemical bond and its oxidation state. This technique has used in the field of forensics [23], biology, and electronics to examine surface chemistry. This technique sustain a UHV environment and this tool is restricted to only solid samples for the investigation of thin film surfaces, semiconductors and metal oxide coatings or DNA/metal film [24–26], pharmaceutical materials (dry powder) [27–29], biosensors [30], and perovskites [31].

Recently, Zhang et al. published a review of factors affecting the photochemical and thermal stability of perovskite materials [32]. However, in solar cells (perovskites) for understanding of technique of thermal and photochemical degradation on level of electronic and atomic, more knowledge of chemical bonding and electronic structure of perovskites is required [33]. To analyze properties, XPS looks to be an appropriate and well organized tool because this technique directly allows the analysis of atom's local chemical bonding as well as total density of all occupied states compared with calculation of band structure. The study of XPS uses measurements of core level to analyze or study C-N bond breaking, $PbBr_2$, PbO, and PbI_2 phase formation under light exposure and annealing (thermal) process of perovskites [34].

This chapter focuses on the merits of a combinatorial approach for material analysis using XPS (lab-based) and their surface technique. The chapter depicts the technique principle in brief, the requirements of the sample, and the conventional strategy of measurements. The feasible information sources like chemical bonding, lateral distribution, depth, and elements are described along with the summary of quantification principles. The technique deals with various applications involving insulators, metals, polymers, and semiconductors to show the method's capabilities over an extensive area of research problems. A comparison of this XPS tool with another complementary analytical method is also summarised. The development in various technologies like nanotechnology, medicine, biotechnology, electronics, and polymers are all interested in surface-related techniques showing sustained concern for the XPS technique in the coming future. The XPS study of various calcium-based natural materials and their applications is done in this chapter.

8.2.1 BASIC PRINCIPLE AND WORKING

XPS is a spectroscopic approach which is based on the photoelectric effect. Its development can be ascribed to the pioneering work of Dr. Kai Siegbahn and his group at Uppsala university, Sweden, in the mid-1960s. In appreciation of his contributions, Dr. Siegbahn was awarded by the Nobel Prize (1981) [20].

The surface of a solid varies with respect to its chemical composition and physical properties compared to its interior. To analyze the surface, beam of X-ray light is utilized to ionize atoms present. The energy of the light utilized must be enough for the ionization of electrons, specifically from the atom's outermost valence shell.

When photons with higher energy ($h\nu$) interact with a crystal surface, eject electrons not only from the outer shells but also from deeper levels within the atoms (Figure 8.3). The X-ray region which is characterized by lower energy wavelengths is effective in knocking out core electrons. The analysis of electrons escaping from the surface allows the acquisition of a spectrum. In XPS, the sample is exposed to low-energy X-rays (~ 1.5 keV) causing the photoelectric effect. Thereafter, spectrometer having high resolution records the emitted photoelectron's energy spectrum [35].

Maintaining a high vacuum environment is essential for XPS experiments. The presence of high vacuum facilitates the transport of photoelectrons to the analyzer while minimizing recontamination of samples. Contamination makes up a remarkable thought in XPS due to the technique's surface sensitivity, with the sampling depth typically limited to a few nanometers [36].

The kinetic energies of the ejected photoelectrons directly give rise to the identification of elements within the sample. Moreover, relative element concentration can be decided by analyzing intensities of the emitted photoelectrons [20].

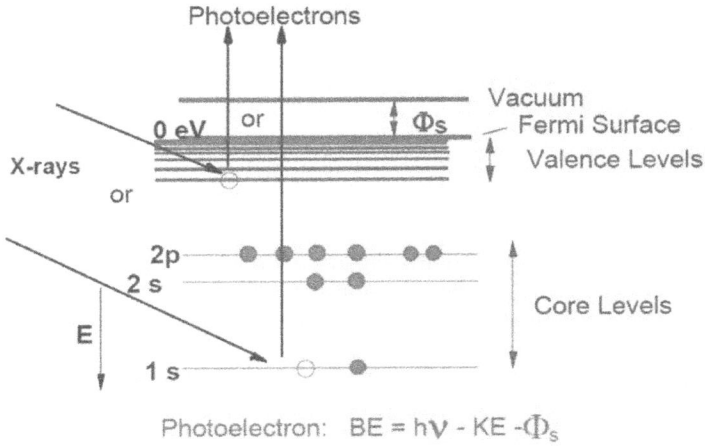

Figure 8.3 XPS basic principle [35].

In XPS, when photons (typically X-rays) of a known energy interact with atom's surface, an electron from K-shell of the atoms is ejected. The K.E. of this ejected electron is then examined in spectroscopic analysis. The resulting spectrum is demonstrated as a plot of B.E. against the counting rate of electrons. B.E. provides a distinctive characteristic for each element and helping in its identification [37]. To ensure surface sensitivity, the X-ray penetration depth into solids is limited to a few microns by adjusting the experimental conditions.

During XPS experiment, interactions between incident photons and the atoms on the surface causing the emission of photoelectrons. The K.E. of ejected electrons is determined by utilizing the following equation [35]:

$$K.E. = h\nu - B.E. - \phi_S \tag{8.1}$$

Here, $h\nu$ represents the photon energy, B.E. corresponds to the atomic orbital binding energy from which the ejection of electron is occurred, and ϕ_S denotes the spectrometer work function.

In an XPS spectrum, innermost orbitals exhibit higher B.E. compared to the outer orbitals. The B.E. of $1s$ orbitals is proportional to the atomic number and it increases accordingly. It has developed empirical formulas to describe the relationship between binding energies and atomic numbers.

A typical spectrum represents a graph or plot depicting the detected electron number as a function of K.E. [20] providing valuable information about the distribution of electron energies within a material. In data analysis, it commonly used the Fermi energy is commonly used as the reference point or natural zero for solids. Each element exists in or on the surface of the analyzed sample constructs a distinctive set of XPS peaks with specific B.E. values. These characteristic peaks directly recognize the elements and correspond to the electronic configurations of electrons in different orbitals, such as $1s$, $2s$, $2p$, $3s$, and so on [35].

The fundamental parts of XPS instrumentation include the following: an excitation source, a sample holder, an analyzer, a detector, and a signal processor. In the time of study of XPS, the sample is exposed to photons with well known energy leading to the occurrence of the photoelectric effect. When bound electron absorbs exposed photon, it absorbs photon's energy and converts it into K.E. As the electron leaves the atom, it overcomes the Coulomb attraction of the nucleus resulting in a decrease in the value of its kinetic energy. The outer orbitals of the atom readjust, reducing in the final state energy. It transferred the extra energy released during this readjustment to the outgoing electron and modified its kinetic energy [38]. It commonly known these outgoing electrons produced near surface, leave the material of sample into vacuum and enter slit of spectrometer's analyzer. The analyzer is able to measure the current of electron which represents electron's number per unit time (function of energy). These plots depicting intensity versus energy as XPS spectra [20].

XPS is commonly used for the investigation of various inorganic compounds, such as polymers, ceramics, metal alloys, glasses, semiconductors, and ion-modified materials. It serves as a useful technique for the analyses of surface chemistry and composition of materials [35].

8.2.2 XPS PLOTS

It is important to have a comprehensive understanding of the mechanism and principles behind this analytical technique for effective interpretation of graphs. XPS has the capability to detect and analyze a wide range of elements excluding hydrogen and helium [35].

As a result, conducting a survey scan is the initial step in most analysis. Each element in the periodic table exhibits various electron states that the X-ray beam can excite. By obtaining an extensive overview of the total domain of elements on the sample's surface, it becomes more feasible to focus on specific XPS spectra during high-resolution investigations.

8.2.2.1 The Horizontal Axis (Peak Position)

During XPS analysis, the horizontal peak position in a spectrum corresponds to the elemental/chemical constitution of the sample and it is expressed in terms of Binding Energy (eV). The B.E. value is determined by calculating the difference in energy between the incident X-ray source energy and the K.E. of the detected photoelectron. Generally, the plot of the horizontal axis is from the highest B.E. to lowest B.E. to accurately represent spectral data [35].

8.2.2.2 The Vertical Axis (Peak Intensity)

During XPS study, the vertical axis represents surface material intensity showing abundance of a particular surface element. This depicts the total photoelectron rate

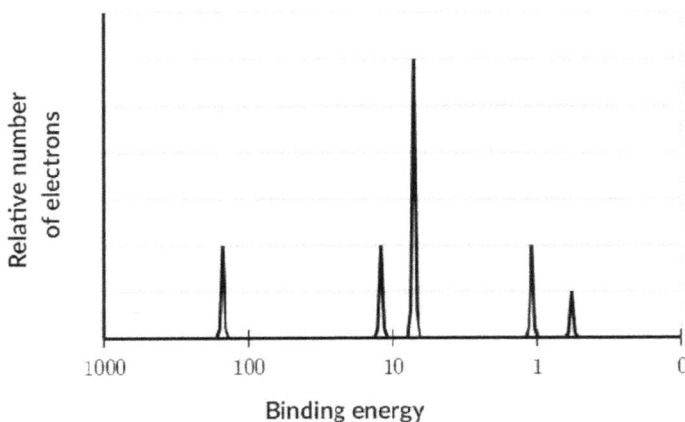

Figure 8.4 Typical XPS spectrum of unknown element [35].

per second which provides a measure of the surface concentration. The graph depicting the photoelectron intensity as a function of binding energy (ranging from approximately 0–1100 eV) is termed as survey scan [35].

8.2.2.3 Identifying element using XPS spectrum

An element (unknown) is investigated and analyzed by using a instrument called photoelectron spectrometer showing in the below spectrum (Figure 8.4). The spectrum exhibits five different peaks, one having B.E. close to 100 units, two having B.E. around ten units, and two having B.E. near one unit. The relative intensities of the peaks from left to right are 2x, 2x, 6x, 2x, and 1x [37].

The XPS survey scan spectrum reveals five peaks, correspond to electrons in the five adjacent subshells near nucleus ($1s$, $2s$, $2p$, $3s$, and $3p$). The peak located at the highest B.E. (leftmost peak) is related to the $1s$ subshell, whereas the peak at the lowest B.E. (rightmost peak) related to the $3p$ subshell. The $3p$ peak demonstrates half the intensity of the $1s$, $2s$, and $3s$ peaks, suggesting the existence of a lone pair electron in the $3p$ subshell. Based on this analysis, it can be concluded that the element which is unidentified is Al [35].

8.2.3 APPLICATIONS OF XPS

In the area of materials science, XPS serves as a valuable tool for obtaining crucial information regarding the constitution, surface chemical state and interface chemical state. Additionally, XPS can give insights into several other factors, including determining the surface's elemental constitution (within 0–10 nm), analyzing the surface constitution of catalysts, establishing the empirical formula of materials, identifying elements which corrupt a surface, examining the electronic state of element present on surface, assessing the consistency of elemental constitution over the top surface

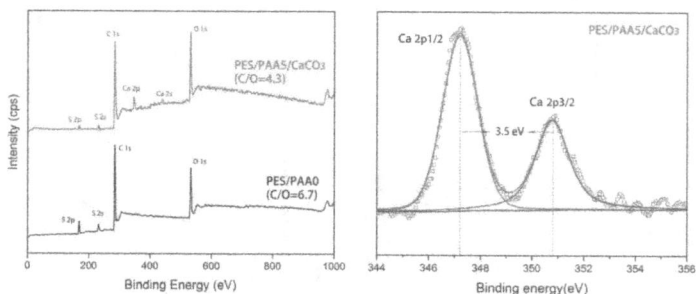

Figure 8.5 XPS wide scan and Ca $2p$ core level spectrum of PES/PAA0 and PES/PAA5/ CaCO$_3$ membranes [39]. (Adapted with permission under CC-BY from Liu *et al. Sci. Rep.* 6:19593, 2016. Copyright 2016 Nature Publishing Group.)

through techniques such as mapping, investigating the uniformity of elemental constitution (depth profiling) [38].

8.2.3.1   Thin film

Liu et al. presented the progress made in developing a polyethersulfone (PES)-based substrate for the study of thin film composite (TFC) forward osmosis (FO) membranes, which includes a hydrophilic mineral (CaCO$_3$) coating and they confirmed CaCO$_3$ presence by XPS [39]. Figure 8.5 depicts the wide scan of XPS and the Ca spectra of the membranes. The observed extra peaks in the PES/PAA/CaCO$_3$ membrane spectrum represent Ca element. The binding energies of 346.8 and 350.3 eV noticed in the Ca $2p$ region orient exactly with the expected Ca $2p_{1/2}$ and Ca $2p_{3/2}$ which depicting the presence of characteristic peaks. Additionally, the survey spectrum of PES/PAA/CaCO$_3$ exhibit an increment in oxygen content and a decrement in C/O ratio in comparison to PES/PAA0 as shown in Figure 8.5 which result agrees with the outcome of CaCO$_3$ deposition.

8.2.3.2   Dental Medicine

Clichici et al. performed XPS analysis for the identification of ions in the bioactive glasses with boron and vanadium (BG2) constitution for the application in dentistry [40]. Figure 8.6 represents F $1s$, Ca $2p$, V $2s$, and B $1s$ core shell XPS spectra of BG2 sample. The peak observed at 632.67 eV in the V $2s$ region suggests an association with V$_2$O$_5$. While the Ca $2p$ line provides limited information regarding chemistry, the presence of the $2p$ (3/2) line at 349.5 eV can be ascribed to CaF$_2$. However, the shift toward higher binding energy implies the existence of Ca-OH bonds near the sample's surface.

8.2.3.3   Biomedical and pharmaceutical

Bolli et al. investigated calcium carbonate functionalized with hydroxyapatite (HA-FCC) matrix comprising Ag nanoparticles materials for application of biomedical

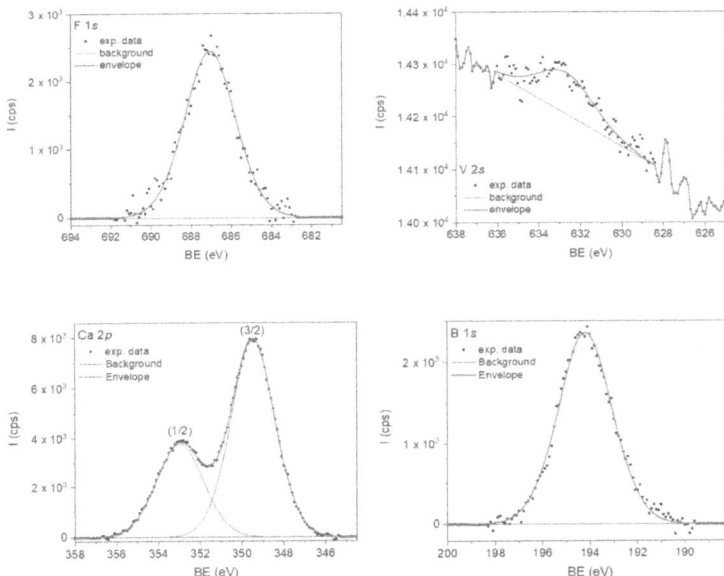

Figure 8.6 XPS spectra of F 1s, V 2s, Ca 2p, and B 1s core-level lines corresponding to sample BG2 [40]. (Adapted with permission under CC-BY from Clichici *et al.* *Materials*. 15:9060. 2022. Copyright 2022 MDPI.)

and pharmaceutical significance. They showed survey spectrum of these materials in Figure 8.7 and observed that evaluated atomic ratios are in close agreement with their stoichiometric compound. The shifting in higher binding energy side by 23 ± 1 and 36 ± 1 eV is presented in all XPS spectra of Figure 8.8 [41].

8.2.3.4 Floor Designing

Chen et al. synthesized nanolignocellulose (NLC) composite materials for application in floor designing [42]. Figure 8.9 shows wide survey spectrum and other XPS spectra of NLC and NLC/CaCO$_3$/PMMA materials ((poly (methyl methacrylate)). They observed extra peak of Ca 2p and Ca 2s of CaCO$_3$ in NLC/CaCO$_3$/PMMA when compared survey spectra. In Figure 8.9(b), Ca $2p_{3/2}$ and Ca $2p_{1/2}$ peaks are located at 347.07 and 350.6 eV and confirmed the existence of CaCO$_3$ in this composite material.

8.2.4 LIMITATIONS OF XPS ANALYSIS

While XPS offers considerable flexibility, it may not be suitable for every material testing. Several limitations arise when utilizing XPS for sample analysis. The following are some constraints during XPS investigations:

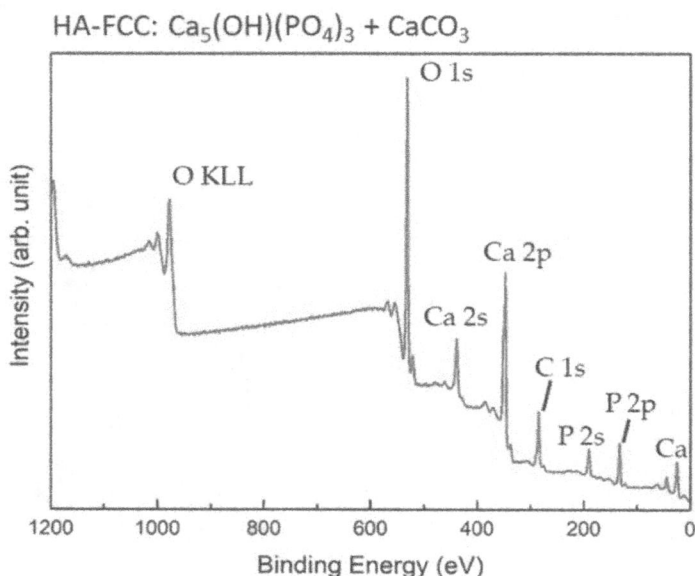

Figure 8.7 Survey spectrum of HA-FCC reference sample. The main peaks of P, Ca, C, and O are indicated [41]. (Adapted with permission under CC-BY from Bolli *et al. Nanomaterials*. 11:2263, 2021. Copyright 2021 MDPI.)

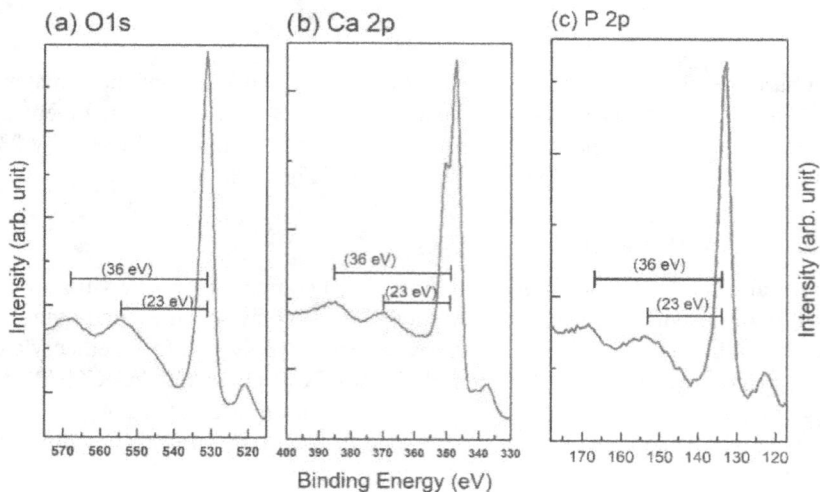

Figure 8.8 XPS spectra of O 1*s* (a), Ca 2*p* (b), and P 2*p* (c) regions. Besides the main photoemission peaks, in all the spectra are clearly visible the shake-up satellites [41]. (Adapted with permission under CC-BY from Bolli *et al. Nanomaterials*. 11:2263, 2021. Copyright 2021 MDPI.)

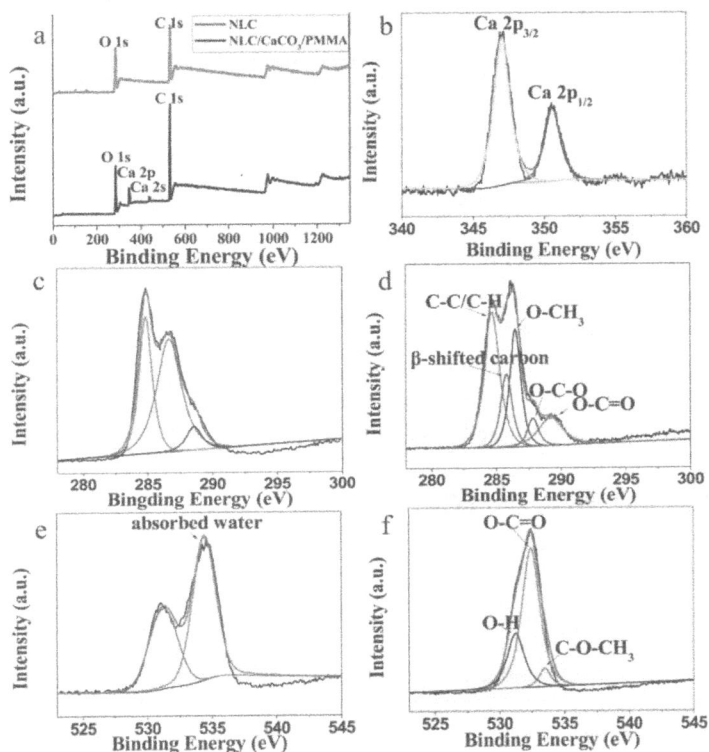

Figure 8.9 XPS spectra of (a) survey spectrum, (b) Ca $2p$ in NLC/CaCO$_3$/PMMA composite and (c,d) C$1s$, (e,f) O$1s$ in pure NLC and NLC/CaCO$_3$/PMMA composite, respectively [42]. (Adapted with permission under CC-BY from Chen *et al. Sci. Rep.* 8:5121, 2018. Copyright 2018 Nature Publishing.)

8.2.4.1 Size of the Samples

The lateral dimensions of the sample must not exceed 25 mm, while the height remain below 1/2 inch (12 mm). It is significant to acknowledge that the sample size can pose limitations, and cutting the sample is not always a viable solution. Sample cutting can introduce additional contamination to the area of interest, resulting in misleading or inaccurate outcomes [43].

8.2.4.2 Challenges of Reproducibility

Like other analysis techniques, XPS exhibits error of approximately 10% in repeated analyses. There can be a deviation of up to 20% between the measured value and the true value of the sample [43].

8.2.4.3 Vacuum Environment

For successful utilization of the XPS technique, it is necessary that the samples are suitable with a high vacuum environment [44]. They need to withstand the conditions of a higher vacuum (below 1×10^{-9} Torr). In case the sample experiences flash out if subjected to vacuum, it will hinder the production of the desired results through the XPS technique [43].

8.2.4.4 Broad Elemental Detection with XPS: Beyond Atomic Number 3

Due to the extremely small orbital size, XPS is unable to detect elements such as hydrogen and helium. The diameter (small) of orbitals remarkably decreases the feasibility of their capture, resulting in an almost negligible catch probability [43].

8.2.4.5 More Limitations

XPS is a technique that requires a spacious analysis area, making it a relatively expansive method. The data collection process sometime can be time-consuming, leading to slower rate of overall analysis [43].

8.3 X-RAY EMISSION SPECTROSCOPY

X-ray emission spectroscopy (XES) and X-ray absorption spectroscopy (XAS) techniques are closely related and provide complementary information about the electronic structure of compounds [45]. In XAS, incident X-ray photons are used to excite electrons in core shells to higher energy empty electronic states through absorption. By studying the absorbed X-ray photons and the corresponding energy of the core electrons, we can obtain qualitative information about the state of the atom/ion present in the sample.

During XAS, the excitation of a core electron to a higher energy level creates a hole at a lower energy level. An electron from a higher orbital subsequently filled this hole, resulting in the emission of radiation corresponding to the energy difference between the initial and final states of the de-excited electron. In XES, we analyze this emitted radiation as a function of energy to obtain qualitative information about the elements present in the sample.

XAS primarily provides insights into the density of unoccupied electronic states, while XES is more closely associated with the density of occupied states. Together, these techniques offer valuable knowledge about the electronic properties of compounds, enabling researchers to investigate and understand their structure and behavior more comprehensively.

Figure 8.10 depicts the X-ray absorption and emission spectra, both plotted on the same energy scale. These spectra adhere to the law of conservation of energy. The two spectra only overlap at the spectral tails, as X-ray emission spectroscopy (XES) provides information about the highest energy occupied states, while X-ray absorption spectroscopy (XAS) begins with the lowest energy unoccupied state [46].

Resonant inelastic X-ray scattering (RIXS), also known as resonant XES, is a spectroscopic technique that combines both XAS and XES. Gautam et al. [48] studied the electronic structure of Co-doped ZnO ($Zn_{1-x}Co_xO$) thin films using O K-edge and Co $L_{3,2}$ edges using near-edge X-ray absorption fine structure (NEXAFS)

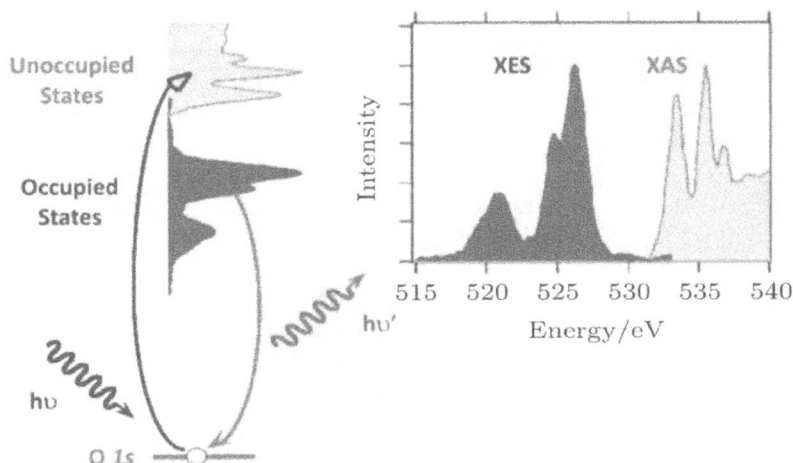

Figure 8.10 A schematic representation of soft X-ray absorption spectroscopy (XAS) and soft-X-ray emission spectroscopy (XES) processes. Soft X-rays cause an electron to be excited to an unoccupied molecular orbital (MO). When another electron fills the resulting hole, X-rays of a different wavelength are emitted [46, 47].

and Resonant inelastic X-ray scattering (RIXS). In this work, RIXS measurements comes out to be an effective way to find out the local electronic structure around Co-dopants. In this work, the Co-dopants exists in two states, one is in the interstitial sites and another in substitutional sites of Zn-atoms, probes by O K-edge of NEXAFS. Through RIXS measurements, Gautam and his coworkers, had clarified whether the minority interstitial Co-dopants, had any interactions with substitutional Co-dopants or not [48].

One advantage that XES has over XAS is that, it does not require high-resolution or tunable incident X-ray photons. In contrast, XAS necessitates radiation with good resolution and tunable energy. Consequently, XES can be performed using X-ray tubes and was conducted even before synchrotron facilities became available to the scientific community [49–51].

8.3.1 BASIC PRINCIPLE AND WORKING

The fundamental principle of X-ray emission spectroscopy (XES) involves the excitation of core electrons to higher energy levels by radiation with sufficient energy to excite the strongly bonded core electron. High-energy X-rays or gamma-rays are commonly used for this excitation, similar to X-ray absorption spectroscopy (XAS), resulting in the creation of a core hole. However, the electrons in higher energy states are relatively unstable and tend to return to their original levels through a process called relaxation. It is important to note that the core hole formed due to excitation lasts only for a very short duration of approximately 10^{-15} s [45].

During the relaxation process, electrons from higher orbitals de-excite to lower energy levels to fill the core hole created by the excitation process. As a result of this de-excitation, X-ray photons are emitted through fluorescence. The energies of these emitted X-rays are characteristic of the nature of the element present in the sample. For example, if the incident X-ray excites the $1s$ core shell (K-shell) electron, the observed fluorescence is called K-fluorescence. Similarly, excitation from $2s$ or $2p$ orbitals (L-shell) leads to L-fluorescence, and so on. The emission lines are also classified as α, β, and γ lines, depending on their relative intensities. Figure 8.11 illustrates these lines, with α-lines being the most intense, followed by the subsequent lines in decreasing order of intensities. The K_α fluorescence line, resulting from the de-excitation of a $2p$ electron to a $1s$ hole, is particularly prominent [45]. These transitions adhere to the same selection rules as other electronic spectroscopy, requiring a change in symmetry between the initial and final orbitals, specifically from gerade to ungerade or vice versa. In XAS and XES, electronic states with p-symmetry (spherical) regarding the metal center are typically investigated.

Every element has distinct energy levels, which means that X-rays emitted during transitions between these fixed energy levels will produce unique emission spectra. During the de-excitation process, multiple electron levels or orbitals are involved due to the presence of multiple electrons in an element. Consequently, different fluorescence emission lines are observed. Additionally, spin-orbit interactions of electrons and electron–electron interactions within the valence and core shells, or between them, play a significant role in determining the intensity of fluorescent emission lines and contribute to the fine structure of an element [45].

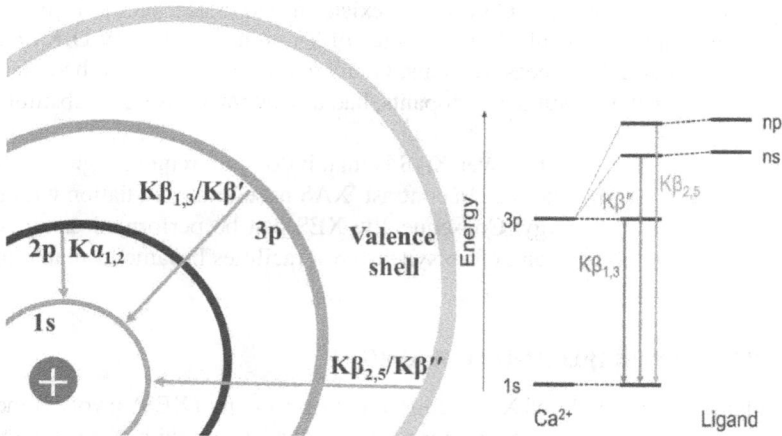

Figure 8.11 Schematic view of various K-emission lines obtained in XES. The transitions from $2p \rightarrow 1s$ results $K_{\alpha_{1,2}}$ lines, $3p \rightarrow 1s$ results $K_{\beta_{1,3}}$, $K_{\beta'}$ emission lines and valence band $\rightarrow 1s$ results $K_{\beta_{2,5}}$ and $K_{\beta''}$ emission lines. In the transitions from valence band$\rightarrow 1s$ orbital, the ligands attached with central atom also plays an important role, as they too a part of valence band.

Spin-orbit interactions are stronger for the K_α line ($2p \rightarrow 1s$ transition) but relatively weaker for K_β lines (transitions involving 3p orbitals or higher to 1s). The $2p$ orbitals split into two orbitals with different energies due to strong spin-orbit coupling, resulting in two emission peaks: K_{α_1} and K_{α_2}. The $3p$ and $3d$ orbitals experience strong inter-electron repulsion, leading to separate $K_{\beta_{1,3}}$ and $K_{\beta'}$ emission lines. Weaker satellite peaks, such as $K_{\beta_{2,5}}$ and $K_{\beta''}$, arise from transitions between valence orbitals and the metal core 1s shell. The K_α and $K_{\beta_{1,3}}$, $K_{\beta'}$ emission lines provide information about the chemical sensitivity arising from electron-electron interactions. The outer 3p electrons interact more with valence electrons through electron spin compared with inner $2p$ electrons. As a result, the $K_{\beta_{1,3}}$, $K_{\beta'}$ emission lines offer a clearer picture of the chemical state of the atom/ion in the sample [45,52].

The sensitivity of the K_α and K_β emission lines to the total spin in the valence shell indirectly provides information about the oxidation state of the element. Unlike electron spin resonance, XES does not exhibit $K_{\alpha,\beta}$ spin species, making it easier to correlate the number of unpaired spins with the position of the $K_{\beta_{1,3}}$ emission line [45,53]. Once the spin state (high spin or low spin) of the element is determined from the position of the $K_{\beta_{1,3}}$ emission line, one can interpret the element's oxidation state [54]. Therefore, any change in the $K_{\beta_{1,3}}$ emission line can be understood as a change in the element's oxidation state. The $K_{\beta_{2,5}}$ and $K_{\beta''}$ emission lines, resulting from valence-to-core transitions, provide information about the orbitals participating in the chemical bond and any changes in the local structure around the element [45]. Similar to X-ray absorption near-edge structure (XANES), a larger peak shift in the $K_{\beta_{2,5}}$ and $K_{\beta''}$ emission lines indicates a greater change in the element's oxidation state. However, these peak shifts occur in opposite directions due to the absorption and emission phenomena in XANES and XES, respectively. The $K_{\beta''}$ emission lines, also known as "crossover" transitions, provide details about the type of ligand and the bond length between the central metal atom and the ligand [55]. In Figure 8.11, the $K_{\beta''}$ valence-to-core emission peaks result from the transition of O 2s to the Mn 1s orbital. Bergmann et al. [55] demonstrated that if the intensity of the $K_{\beta''}$ peak is normalized to the main $K_{\beta_{1,3}}$ region and the number of oxygen ligands, the $K_{\beta''}$ peak increases logarithmically with the Mn-O bond distance. Furthermore, ligands such as F, N, and C have decreasing binding energies for 2s orbitals, respectively, resulting in even larger energy shifts [56].

Valence-to-core emission spectra can provide information about the ionization energies of ligands attached to a central atom and reveal details about the occupied valence molecular orbitals of the ligands. Valence-to-core emission spectra have an advantage over extended X-ray absorption fine structure (EXAFS) in identifying lighter atoms attached to transition metals that may not be resolved in EXAFS [57]. EXAFS can not distinguish between the atoms having almost similar atomic numbers, like C, N, and O, in such a situation, valence to core X-ray emission spectroscopy comes to play [57]. The analysis of fluorescence lines coming from the valence electronic levels, linked with the chemical environment. Even though the ligands have almost similar atomic numbers, but their valence electronic states and there chemical environments differ, which can be easily detected with the emission lines [58,59].

However, valence-to-core emission spectra have the limitation of not being able to re-solve the symmetry of the coordination environment around the central atom, which can be determined more easily through the pre-edge region of XAS.

Directly determining the exact nature of the transitions involved in this process is challenging. However, we can approximate information about these transitions by considering the energy of absorbed and emitted photons. According to the law of conservation of energy, the energy of the transition is the difference between the energy of the absorbed photon and the energy of the emitted photon(s). Different atoms have distinct chemical environments, leading to different energy levels. At the core of XES lies the measurement of changes in the K emission spectrum for atoms with varying chemical surroundings.

Some key features of XES include:

1. XES studies do not require synchrotron radiation with monochromatic and tunable radiation. An analyzer with an energy resolution of 1–2 eV is suffi-cient for studying 3d transition metals [60].
2. The intense spectral region of $K_{\alpha_{1,2}}$ is not highly sensitive to changes in the local structure.
3. The $K_{\beta_{1,3}}/K_{\beta'}$ regions reveal changes in spin and oxidation states of $3d$ transition metals such as Mn, Fe, and Co [61, 62].
4. The spectral region of $K_{\beta_{2,5}}/K_{\beta''}$ in the emission spectra is highly sensitive to electronic structure and local coordination. The $K_{\beta''}$ emission peak pro-vides information about the nature of the ligand and its interatomic distance from the central atom.

8.3.2 XES SPECTRA OF CALCIUM-BASED PRODUCTS

In nature, calcium is predominantly found in the stable form of Ca^{2+}. The d^0 con-figuration of calcium makes it challenging to study using spectroscopic techniques such as UV-visible spectroscopy or Electron Paramagnetic Resonance (EPR) spec-troscopy [63]. However, in XAS and XES techniques, calcium is commonly investi-gated through its K-edge. The K-edge in XES of calcium provides information about its coordination environment, including the types and number of ligands, oxidation state, interatomic bond lengths, and more [63].

In XES at the K-edge of calcium, the $1s$ electron is ionized through the absorption of X-rays, leaving behind a core hole. This core hole is then filled by other electrons from higher energy shells, resulting in the emission of a second X-ray. For calcium, the transition of the $3p$ electron to the core $1s$ hole leads to the emission of an intense fluorescent X-ray, which manifests as a prominent $K_{\beta_{1,3}}$ emission line in the spec-trum. Due to the absence of electrons in the d orbitals of calcium, the $K_{\beta'}$ feature, which arises from the transition from the $3d$ orbital to the core $1s$ hole, is absent. Additionally, weak satellite peaks such as $K_{\beta''}$ and $K_{\beta_{2,5}}$ are observed at higher en-ergies. These peaks result from transitions involving the valences s and p orbitals of the ligands attached to calcium.

Figure 8.12 showcases the valence-to-core emission spectra of various compounds, including CaF_2, $CaCl_2$, $CaBr_2$, CaI_2, CaO, $Ca(OH)_2$, $CaCO_3$, LMn_3CaO_4 $(ON_4O)(OAc)$ denoted as **1**, $[LMn_3CaO_3(OH)(ON_4O)(OAc)][OTf]$ denoted as **1H**, and $LMn_3CaO_4(OAc)_3(DMF)$ denoted as **2** [63]. Triflate (OTf) is used as a ligand, OAc represents acetate, and X represents any other ligand. The observed spectra are displayed on the left side, while the calculated valence-to-core emission spectra of the Ca atom for these compounds are shown on the right side. The calculated spectra were obtained using density functional theory (DFT) calculations with the ORCA 4.1 suite [64].

Broadly, the emission spectra exhibit two main regions: (1) between 4005 and 4024 eV, and (2) 4024 and 4038 eV.

The first region comprises strong $K_{\beta_{1,3}}$ mainlines, corresponding to Ca $3p \rightarrow 1s$ transitions at approximately 4012 eV [65]. Weaker $K_{\beta''}$ transitions are also observed, which might originate from transitions involving multiple excited states and the donor orbitals primarily with ligands of s-character [66]. Unlike transition metal valence-to-core emission spectra, calcium does not possess d-electrons. As a result, there is no exchange with p-holes of the final states, and therefore no $K_{\beta'}$ emission lines are observed [63]. Notably, the intensity of the Ca $K_{\beta_{2,5}}$ spectral feature is nearly double compared to the same emission line in transition metal complexes [67–69]. In transition metal complex XES, a clear distinction between $K_{\beta''}$ and $K_{\beta_{1,3}}$ emission lines is typically observed, and the difference between these two lines provides information about the chemical environment surrounding the transition metal ion [8, 69]. However, in the case of calcium compounds, it is challenging to discern this distinction as the $K_{\beta''}$ lines are not well resolved from the mainlines [63]. To overcome this challenge, interpretations of the emission spectra often rely on the $K_{\beta_{2,5}}$ emission lines.

Figure 8.13 represents the $K_{\beta_{2,5}}$ emission spectra of Calcium halides. The crystal environment of all calcium halides are symmetric and emission peak shown in Figure 8.14 are the result of a transition from the halide p-orbitals to the core $1s$ hole of calcium. In this figure the peak shift from fluorine to iodine compounds attributes to

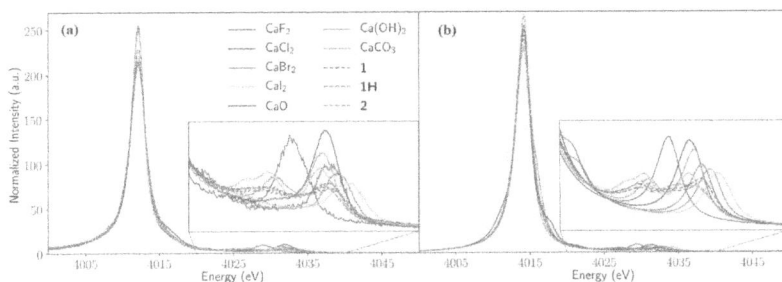

Figure 8.12 All observed (a) and calculated (b) calcium valence to core XES spectra, with the $K_{\beta_{2,5}}$ regions inset [63]. (Adapted with permission under CC-BY from Mathe *et al. Inorganic Chem.* 58: 16292–16301, 2019. Copyright 2019 ACS Publications.)

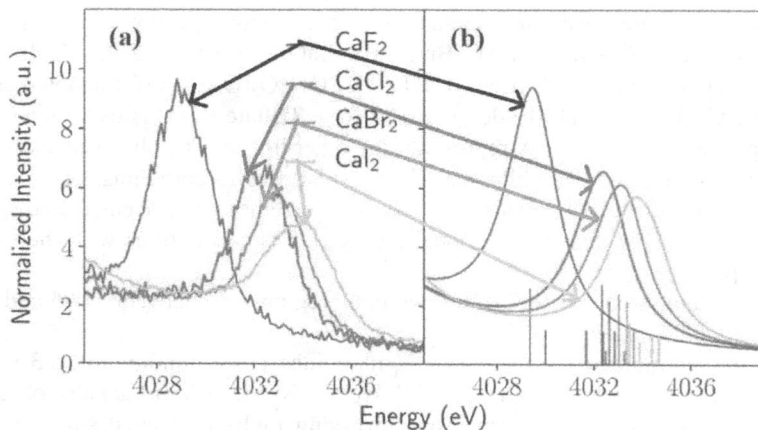

Figure 8.13 Observed (a) and calculated (b) $K_{\beta_{2,5}}$ spectra of the calcium halides, with individual calculated transitions plotted as sticks. [63]. (Adapted with permission under CC-BY from Mathe *et al. Inorganic chemistry* 58(2019) 16292–16301, Copyright 2019 ACS Publications.)

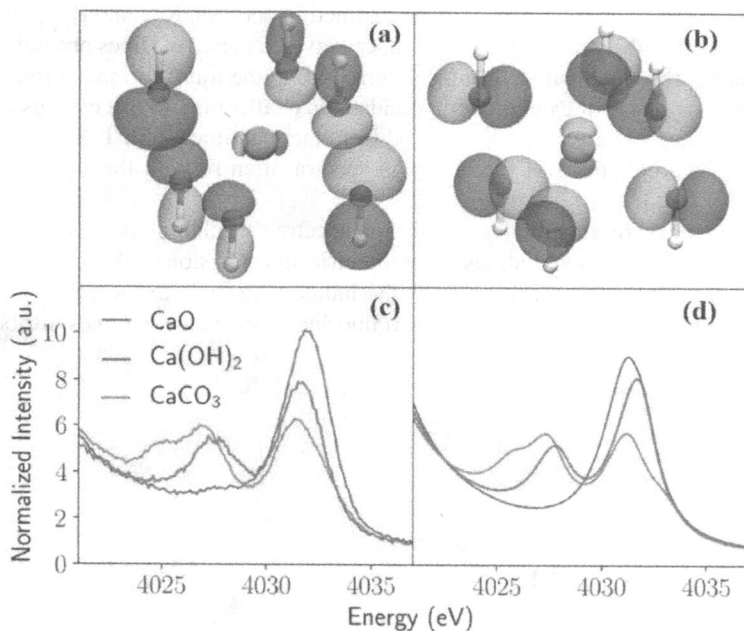

Figure 8.14 Representative donor orbital plots for the low-energy (a) and high-energy (b) features of Ca(OH)$_2$. Observed (c) and calculated (d) spectra of the oxygen-containing calcium salts [63]. (Adapted with permission under CC-BY from Mathe *et al. Inorganic Chem.* 58:16292–16301, 2019. Copyright 2019 ACS Publications.)

the decrease in electronegativity of the halogen atom, because the highest occupied molecular orbitals(HOMO's) in this case is halogen's donor orbitals, and more the tightly these HOMO's with p-orbital character are attached to the lighter halogen atom, they lie closer to the core $1s$ hole of the calcium, thus have emission lines relatively at lower energy [63].

Figure 8.14 represents the observed and calculated $K_{\beta_{2,5}}$ emission spectra of different oxygen coordinated calcium salts along with their representative donor orbital plots. The $K_{\beta_{2,5}}$ emission lines are as a result of transitions from valence orbitals of oxygen atoms to the core $1s$ hole of calcium atom. A clear peak separation, along with small shoulders for $CaCO_3$, attributes to the transition from σ^*, π, π, and π^* molecular orbitals of a carbonate ligand to the core hole of calcium atom. Similarly, splitting of emission peak with no shoulders, is also observed for $Ca(OH)_2$. To explain this, let us consider the O-H bond axis is along z-axis. The emission peak observed at higher energy is due to the transition from two oxygen donor atoms p_x and p_y, while the lower energy peak is because of a transition from a single O p_z donor orbital, which is also overlapped with the hydrogen atom [63]. Since, transitions from two orbitals at higher energy, compared to a single orbital transition at lower energy, thus the observed intensity ratio will be 2:1, because all the three sets of p-orbitals have similar occupation and orientation along Ca atom, therefore have equal probability of transitions. In the case of CaO, there is no hydrogen atom involved and thus all the $3p$-orbitals are of equal energy and therefore will give transition at equal energy, therefore no peak splitting is observed in this case [63]. Mathe et al. [63] has also discussed the valence to core XES of other calcium based products, and can be helpful for further reading.

8.4 CONCLUSION

Calcium has vast area of applications whether it is in industries and in the medical field, and it depends on Ca^{+2} ion interaction found in proteins, leaves, teeth, etc. Calcium phosphate which is a biocompatible material used for formation of hard tissues, capping agent, cleft palate, apical barrier, bone defects, etc. Not only in medical but its compound like calcium peroxide used in wastewater treatment, soil remediation, groundwater and also used in dye removal, petroleum hydrocarbons [71]. Calcium carbonate whisker increased the physical and chemical properties that improve the heat resistance property and enhanced the cement-based composites [72]. Nowadays, calcium orthophosphate [73] has been using in artificial replacements for fracture knees, decayed teeth, for stablilization of jawbone, spinal fusion and bone fillers etc. These extensive application has been developed with the wide knowledge of calcium behaviour, its environment, its chemical properties and many more could be found out with the potential of XAS, XPS, and XES techniques.

In conclusion, XAS, XPS, and XES of calcium-based natural products offers valuable insights into the electronic structure and chemical environment of calcium atoms within these complex biological systems. By utilizing the characteristic these spectroscopic techniques, researchers can identify and analyze the local coordination of calcium ions, as well as the bonding interactions with surrounding ligands. This analytical technique has proven to be a powerful tool for studying the composition and reactivity of calcium-containing biomolecules, including proteins, enzymes, and other essential cellular components. XAS enables the study of calcium's local coordination and electronic state, shedding light on its role in biological systems. XPS allows for the precise identification of calcium oxidation states and surface chemical properties, providing valuable information about the interactions with ligands. XES further contributes to understanding the electronic transitions and valence states of calcium, complementing the knowledge gained from XAS and XPS. Together, these spectroscopic methods offer a holistic understanding of the physicochemical characteristics of calcium-based natural products, from their bulk properties to their surface interactions. By unraveling the molecular intricacies of these biomolecules, this integrated approach holds significant promise for advancing our knowledge in diverse fields, including bioinorganic chemistry, pharmacology, and material science. The continued application of XAS, XPS, and XES in calcium-based natural product research opens new avenues for innovative solutions in healthcare, agriculture, and environmental sustainability. The non-destructive nature of X-ray emission spectroscopy enables the investigation of natural products in their native states, avoiding potential alterations or artifacts. As our understanding of the roles of calcium in biological processes continues to develop, these X-ray based spectroscopy techniques, remains a crucial tool in unraveling the intricate molecular mechanisms underpinning calcium-based natural products' functions, with wide-ranging implications for health, nutrition, and pharmaceutical research.

8.5 ACKNOWLEDGMENTS

SG is thankful to the UGC-DAE CSR for providing the financial support from through a Collaborative Research Scheme (CRS) project number CRS-ISUM-53/CRS-336.

REFERENCES

1. L. Singh, D. Ali, I. Tyagi, U. Sharma, R. Singh, and P. Hou. Durability studies of nano-engineered fly ash concrete. *Construction and Building Materials*, **194**:205–215, 2019.
2. R. A. Scott. [23] measurement of metal-ligand distances by EXAFS. In Methods in enzymology, volume 117, pages 414–459. Elsevier, 1985
3. D. Koningsberger. Principles, applications, techniques of EXAFS, SEXAFS and XANES. *X-ray Absorption*, ISSN - 21876886, 1988.
4. P. Lee, P. Citrin, P. Eisenberger, and B. Kincaid. Extended X-ray absorption fine structureits strengths and limitations as a structural tool. *Reviews of Modern Physics*, **53**(4):769, 1981.
5. H. Iwayama, M. Nagasaka, I. Inoue, S. Owada, M. Yabashi, and J. R. Harries. Demonstration of transmission mode soft X-ray nexafs using third-and fifth-order harmonics of fel radiation at sacla bl1. *Applied Sciences*, **10**(21):7852, 2020.
6. D. Arvanitis, U. Dobler, L. Wenzel, K. Baberschke, and J. Stöhr. A new technique for submonolayer nexafs: Fluorescence yield at the carbon K-edge. *Le Journal de Physique Colloques*, **47**(C8):C8–173, 1986.

7. A. Nefedov and C. Wöll. Advanced applications of NEXAFS spectroscopy for functionalized surfaces. *Surface Science Techniques*, 277–303, 2013.

8. K. M. Lancaster, M. Roemelt, P. Ettenhuber, Y. Hu, M. W. Ribbe, F. Neese, U. Bergmann, and S. DeBeer. X-ray emission spectroscopy evidences a central carbon in the nitrogenase iron-molybdenum cofactor. *Science*, 334(6058):974–977, 2011.

9. G. Cressey, R. Schmid, and B. Wood. Thermodynamic properties of almandine-grossular garnet solid solutions. *Contributions to Mineralogy and Petrology*, 67(4):397–404, 1978.

10. R. Newton, T. Charlu, and O. Kleppa. Thermochemistry of high pressure garnets and clinopyroxenes in the system CaO-MgO-Al_2O_3-SiO_2. *Geochimica et Cosmochimica Acta*, 41(3):369–377, 1977.

11. P. Lee and G Beni. New method for the calculation of atomic phase shifts: Application to extended X-ray absorption fine structure (EXAFS) in molecules and crystals. *Physical Review B*, 15(6):2862, 1977.

12. C. Natoli and M. Benfatto. A unifying scheme of interpretation of X-ray absorption spectra based on the multiple scattering theory. *Le Journal de Physique Colloques*, 47(C8):C8–11, 1986.

13. J. Chaboy and S. Quartieri. X-ray absorption at the Ca-K-edge in natural-garnet solid solutions: A full-multiple-scattering investigation. *Physical Review B*, 52(9):6349, 1995.

14. K. O. Andersson, M. K. Tighe, C. N. Guppy, P. J. Milham, T. I. McLaren, C. R. Schefe, and E. Lombi. XANES demonstrates the release of calcium phosphates from alkaline vertisols to moderately acidified solution. *Environmental Science & Technology*, 50(8):4229–4237, 2016.

15. E. Weyers, D. G. Strawn, D. Peak, A. D. Moore, L. L. Baker, and B. Cade-Menun. Phosphorus speciation in calcareous soils following annual dairy manure amendments. *Soil Science Society of America Journal*, 80(6):1531–1542, 2016.

16. J. Prietzel, W. Klysubun, and L. C. Colocho Hurtarte. The fate of calcium in temperate forest soils: A Ca K-edge XANES study. *Biogeochemistry*, 152:195–222, 2021.

17. D. Rieger, F. Himpsel, U. O. Karlsson, F. McFeely, J. Morar, and J. Yarmoff. Electronic structure of the CaF 2/Si (111) interface. *Physical Review B*, 34(10):7295, 1986.

18. S. Naftel, T. Sham, Y. Yiu, and B. Yates. Calcium l-edge xanes study of some calcium compounds. *Journal of Synchrotron Radiation*, 8(2):255–257, 2001.

19. M. E. Fleet and X. Liu. Calcium L 2, 3-edge XANES of carbonates, carbonate apatite, and oldhamite (CaS). *American Mineralogist*, 94(8-9):1235–1241, 2009.

20. G. Greczynski and L. Hultman. X-ray photoelectron spectroscopy: Towards reliable binding energy referencing. *Progress in Materials Science*, 107:100591, 2020.

21. J. Metson, M. Hyland, A. Gillespie, and M. Hemmingsen-Jensen. X-ray photoelectron spectroscopy applications to corrosion and adhesion at metal oxide surfaces. *Colloids and Surfaces A: Physicochemical and Engineering Aspects*, 93:173–180, 1994.

22. X. Zhang and M. Cresswell. Chapter 3–materials characterization of inorganic controlled release. *Inorganic Controlled Release Technology*, pages 57–91, 2016.

23. J. F. Watts. The potential for the application of X-ray photoelectron spectroscopy in forensic science. *Surface and Interface Analysis*: An International Journal devoted to the development and application of techniques for the analysis of surfaces, interfaces and thin films, 42(5):358–362, 2010.

24. S. Geng, S. Zhang, and H. Onishi. Xps applications in thin films research. *Materials Technology*, 17(4):234–240, 2002.

25. Z. Zhou, Y. Zhao, and Z. Cai. Low-temperature growth of zno nanorods on pet fabrics with two-step hydrothermal method. *Applied Surface Science*, **256**(14):4724–4728, 2010.

26. A. C. D. Luca, J. Stevens, S. Schroeder, J.-B. Guilbaud, A. Saiani, S. Downes, and G. Terenghi. Immobilization of cell-binding peptides on poly-ε-caprolactone film surface to biomimic the peripheral nervous system. *Journal of Biomedical Materials Research Part A*, **101**(2):491–501, 2013.

27. M. Morales, M. Ruiz, I. Oliva, M. Oliva, and V. Gallardo. Chemical characterization with xps of the surface of polymer microparticles loaded with morphine. International *Journal of Pharmaceutics*, **333**(1-2):162–166, 2007.

28. E. Celasco, I. Valente, D. L. Marchisio, and A. A. Barresi. Dynamic light scattering and X-ray photoelectron spectroscopy characterization of pegylated polymer nanocarriers: Internal structure and surface properties. *Langmuir*, **30**(28):8326–8335, 2014.

29. J. S. Stevens, S. J. Byard, C. C. Seaton, G. Sadiq, R. J. Davey, and S. L. Schroeder. Crystallography aided by atomic core-level binding energies: Proton transfer versus hydrogen bonding in organic crystal structures. *Angewandte Chemie International Edition*, **50**(42):9916–9918, 2011.

30. A. Griffith, A. Glidle, G. Beamson, and J. M. Cooper. Determination of the biomolecular composition of an enzyme-polymer biosensor. *The Journal of Physical Chemistry B*, **101**(11):2092–2100, 1997.

31. I. S. Zhidkov, D. W. Boukhvalov, A. F. Akbulatov, L. A. Frolova, L. D. Finkelstein, A. I. Kukharenko, S. O. Cholakh, C. Chueh, P. A. Troshin, and E. Z. Kurmaev. XPS spectra as a tool for studying photochemical and thermal degradation in APBX3 hybrid halide perovskites. *Nano Energy*, **79**:105421, 2021.

32. C. Zhang, T. Shen, D. Guo, L. Tang, K. Yang, and H. Deng. Reviewing and understanding the stability mechanism of halide perovskite solar cells. *InfoMat*, **2**(6):1034–1056, 2020.

33. J. Il, J. Choi, M. E. Khan, Z. Hawash, K. Jeong Kim, H. Lee, L. K. Ono, Y. Qi, Y. Kim, and J. Y. Park. Atomic-scale view of stability and degradation of single-crystal MAPBBR 3 surfaces. *Journal of Materials Chemistry A*, **7**(36):20760–20766, 2019.

34. I. S. Zhidkov, A. F. Akbulatov, A. I. Kukharenko, S. O. Cholakh, K. J. Stevenson, P. A. Troshin, and E. Z. Kurmaev. Influence of halide mixing on thermal and photochemical stability of hybrid perovskites: XPS studies. *Mendeleev Communications*, **28**(4):381–383, 2018.

35. R. Sudhanshu. X-ray photoelectron spectroscopy (XPS) technology. *Its a M.Tech. report*. 2020.

36. F. A. Stevie, R. Garcia, J. Shallenberger, J. G. Newman, and C. L. Donley. Sample handling, preparation and mounting for XPS and other surface analytical techniques. *Journal of Vacuum Science & Technology A: Vacuum, Surfaces, and Films*, **38**(6):063202, 2020.

37. G. Greczynski and L. Hultman. A step-by-step guide to perform X-ray photoelectron spectroscopy. *Journal of Applied Physics*, **132**(1):011101, 2022.

38. D. N. G. Krishna and J. Philip. Review on surface-characterization applications of X-ray photoelectron spectroscopy (XPS): Recent developments and challenges. *Applied Surface Science Advances*, **12**:100332, 2022.

39. Q. Liu, J. Li, Z. Zhou, J. Xie, and J. Y. Lee. Hydrophilic mineral coating of membrane substrate for reducing internal concentration polarization (ICP) in forward osmosis. *Scientific Reports*, **6**(1):19593, 2016.

40. A. Clichici, G. A. Filip, M. Achim, I. Baldea, C. Cristea, G. Melinte, O. Pana, L. B. Tudoran, D. Dudea, and R. Stefan. Characterization and in vitro biocompatibility of two new bioglasses for application in dental medicinea preliminary study. *Materials*, 15(24):9060, 2022.

41. E. Bolli, S. Kaciulis, A. Mezzi, V. Ambrogi, M. Nocchetti, L. Latterini, A. D. Michele, and G. Padeletti. Hydroxyapatite functionalized calcium carbonate composites with Ag nanoparticles: An integrated characterization study. *Nanomaterials*, 11(9):2263, 2021.

42. Y. Chen, T. Cai, B. Dang, H. Wang, Y. Xiong, Q. Yao, C. Wang, Q. Sun, and C. Jin. The properties of fibreboard based on nanolignocelluloses/caco3/pmma composite synthesized through mechano-chemical method. *Scientific Reports*, 8(1):5121, 2018.

43. J. D. Andrade. X-ray photoelectron spectroscopy (XPS). Surface and Interfacial Aspects of Biomedical Polymers. *Surface Chemistry and Physics*, 1:105–195, 1985.

44. C. Das, M. Wussler, T. Hellmann, T. Mayer, and W. Jaegermann. In situ XPS study of the surface chemistry of mapi solar cells under operating conditions in vacuum. *Physical Chemistry Chemical Physics*, 20(25):17180–17187, 2018.

45. U. Bergmann and P. Glatzel. X-ray emission spectroscopy. *Photosynthesis Research*, 102:255–266, 2009.

46. W. Yang and R. Qiao. Soft X-ray spectroscopy for probing electronic and chemical states of battery materials. *Chinese Physics B*, 25(1):017104, 2015.

47. C. Dong, J. Guo, Y. Chen, and C. Chang. Soft-X-ray spectroscopy probes nanomaterial-based devices. *Spie Newsroom*, pages 3–6, 2007.

48. S. Gautam, P. Thakur, P. Bazylewski, R. Bauer, A. Singh, J. Kim, M. Subramanian, R. Jayavel, K. Asokan, K. H. Chae, et al. Spectroscopic study of Zn1- XCOXO thin films showing intrinsic ferromagnetism. *Materials Chemistry and Physics*, 140(1):130–134, 2013.

49. A. Meisel, G. Leonhardt, and R. Szargan. X-ray spectra and chemical binding, volume 37. *Springer*, 1989.

50. K. Tsutsumi. The X-ray non-diagram lines kβ' of some compounds of the iron group. *Journal of the Physical Society of Japan*, 14(12):1696–1706, 1959.

51. K. Tsutsumi, H. Nakamori, and K. Ichikawa. X-ray mn k β emission spectra of manganese oxides and manganates. *Physical Review B*, 13(2):929, 1976.

52. G. Peng, F. Degroot, K. Hämäläinen, J. Moore, X. Wang, M. Grush, J. Hastings, D. Siddons, and W. Armstrong. High-resolution manganese X-ray fluorescence spectroscopy. oxidation-state and spin-state sensitivity. *Journal of the American Chemical Society*, 116(7):2914–2920, 1994.

53. P. Glatzel. X-ray fluorescence emission following K capture and 1s photoionization of Mn and Fe in various chemical environments. PhD thesis, Staats-und Universitätsbibliothek Hamburg Carl von Ossietzky, 2001.

54. G. Vankö, T. Neisius, G. Molnar, F. Renz, S. Karpati, A. Shukla, and F. M. D. Groot. Probing the 3D spin momentum with x-ray emission spectroscopy: The case of molecular-spin transitions. *The Journal of Physical Chemistry B*, 110(24):11647–11653, 2006.

55. U. Bergmann, M. M. Grush, C. R. Horne, P. DeMarois, J. E. Penner-Hahn, C. F. Yocum, D. Wright Dubé, W. H. Armstrong, G. Christou, et al. Characterization of the Mn oxidation states in photosystem II by k$_\beta$ X-ray fluorescence spectroscopy. *The Journal of Physical Chemistry B*, 102(42):8350–8352, 1998.

56. V. Stojanoff, K. Hämäläinen, D. Siddons, J. Hastings, L. Berman, S. Cramer, and G. Smith. A high-resolution X-ray fluorescence spectrometer for near-edge absorption studies. *Review of Scientific Instruments*, **63**(1):1125–1127, 1992.

57. E. Gallo and P. Glatzel. Valence to core X-ray emission spectroscopy. *Advanced Materials*, **26**(46):7730–7746, 2014.

58. G. Smolentsev, A. V. Soldatov, J. Messinger, K. Merz, T. Weyhermüller, U. Bergmann, Y. Pushkar, J. Yano, V. K. Yachandra, and P. Glatzel. X-ray emission spectroscopy to study ligand valence orbitals in mn coordination complexes. *Journal of the American Chemical Society*, **131**(36):13161–13167, 2009.

59. M. A. Beckwith, M. Roemelt, M. Collomb, C. DuBoc, T. Weng, U. Bergmann, P. Glatzel, F. Neese, and S. DeBeer. Manganese $k\beta$ X-ray emission spectroscopy as a probe of metal–ligand interactions. *Inorganic Chemistry*, **50**(17):8397–8409, 2011.

60. U. Bergmann, P. Glatzel, F. d. Groot, and S. Cramer. High resolution k capture X-ray fluorescence spectroscopy: A new tool for chemical characterization. *Journal of the American Chemical Society*, **121**(20):4926–4927, 1999.

61. S. A. Pizarro, P. Glatzel, H. Visser, J. H. Robblee, G. Christou, U. Bergmann, and V. K. Yachandra. Mn oxidation states in tri-and tetra-nuclear mn compounds structurally relevant to photosystem ii: Mn k-edge X-ray absorption and $k\beta$ X-ray emission spectroscopy studies. *Physical Chemistry Chemical Physics*, **6**(20):4864–4870, 2004.

62. H. Visser, E. Anxolabéhére-Mallart, U. Bergmann, P. Glatzel, J. H. Robblee, S. P. Cramer, J. Girerd, K. Sauer, M. P. Klein, and V. K. Yachandra. Mn k-edge xanes and $k\beta$ xes studies of two mn- oxo binuclear complexes: investigation of three different oxidation states relevant to the oxygen-evolving complex of photosystem ii. *Journal of the American Chemical Society*, **123**(29):7031–7039, 2001.

63. Z. Mathe, D. A. Pantazis, H. B. Lee, R. Gnewkow, B. E. V. Kuiken, T. Agapie, and S. DeBeer. Calcium valence-to-core X-ray emission spectroscopy: A sensitive probe of oxo protonation in structural models of the oxygen-evolving complex. *Inorganic Chemistry*, **58**(23):16292–16301, 2019.

64. F. Neese. The orca program system. *Wiley Interdisciplinary Reviews: Computational Molecular Science*, **2**(1):73–78, 2012.

65. Y. Ito, T. Tochio, M. Yamashita, S. Fukushima, A. Vlaicu, K. Słabkowska, E. Weder, M. Polasik, K. Sawicka, P. Indelicato, et al. Structure of high-resolution k β 1, 3 X-ray emission spectra for the elements from Ca to Ge. *Physical Review A*, **97**(5):052505, 2018.

66. P. Glatzel and U. Bergmann. High resolution 1s core hole X-ray spectroscopy in 3D transition metal complexeselectronic and structural information. *Coordination Chemistry Reviews*, **249**(1–2):65–95, 2005.

67. J. A. Rees, V. Martin-Diaconescu, J. A. Kovacs, and S. DeBeer. X-ray absorption and emission study of dioxygen activation by a small-molecule manganese complex. *Inorganic Chemistry*, **54**(13):6410–6422, 2015.

68. E. R. Hall, C. J. Pollock, J. Bendix, T. J. Collins, P. Glatzel, and S. DeBeer. Valence-to-core-detected X-ray absorption spectroscopy: Targeting ligand selectivity. *Journal of the American Chemical Society*, **136**(28):10076–10084, 2014.

69. C. J. Pollock, K. Grubel, P. L. Holland, and S. DeBeer. Experimentally quantifying small-molecule bond activation using valence-to-core X-ray emission spectroscopy. *Journal of the American Chemical Society*, **135**(32):11803–11808, 2013.

70. T. G. Spiro. Calcium in biology, volume 6. Wiley-interscience, 1983.

71. S. Lu, X. Zhang, and Y. Xue. Application of calcium peroxide in water and soil treatment: A review. *Journal of Hazardous Materials*, **337**:163–177, 2017.

72. H. Saulat, M. Cao, M. M. Khan, M. Khan, M. M. Khan, and A. Rehman. Preparation and applications of calcium carbonate whisker with a special focus on construction materials. *Construction and Building Materials*, **236**:117613, 2020.

73. S. V. Dorozhkin. Medical application of calcium orthophosphate bioceramics. *Bio*, **1**(1):1–51, 2011

9 Calcium-Based Waste Material for Catalysis

Current State and Perspective

Ritu Gupta and Ashiya Khan
Dr. S.S. Bhatnagar University Institute of Engineering and
Technology, Panjab University, Chandigarh, India

9.1 INTRODUCTION

The development and design of catalytic material made through readily available and inexpensive components on earth has recently attracted renewed interest in the field of catalysis. The technology is more affordable and environmentally friendly by generating the catalyst from waste material in addition to the target products. Where there are already financial and environmental cost involved with their destructions, consuming waste material is quite advantageous. Sustainability of waste material also be considered.

Many researchers have focused on the design and creation of catalytic material made from cheap, abundant and renewable resources found on earth to increase the overall sustainability of catalytic processes and to minimise the cost of catalyst preparation and loss of metal during the reactions [1, 2]. Calcium rich waste material are widely available in the world as a low-cost renewable resource [3]. The use of calcium-based waste material for catalysis has a significant role in various field of applications such as biodiesel production via transesterification of fats or oil (vegetable oil or fats with alcohol in the presence of catalyst), Biomass pyrolysis/gasification, water purification, catalysis and adsorption [4]. The use of biofuel is growing more spread, due to its beneficial characteristics for the environment such as non-toxicity, highly renewable character and low sulfur content. Biodiesel is a possible replacement for diesel fuel made from crude oil [5]. The decomposition of various raw materials used as sources of $CaCO_3$ and the production of CaO depend on the calcination process as shown in Figure 9.1 [6].

DOI: 10.1201/9781003360599-9

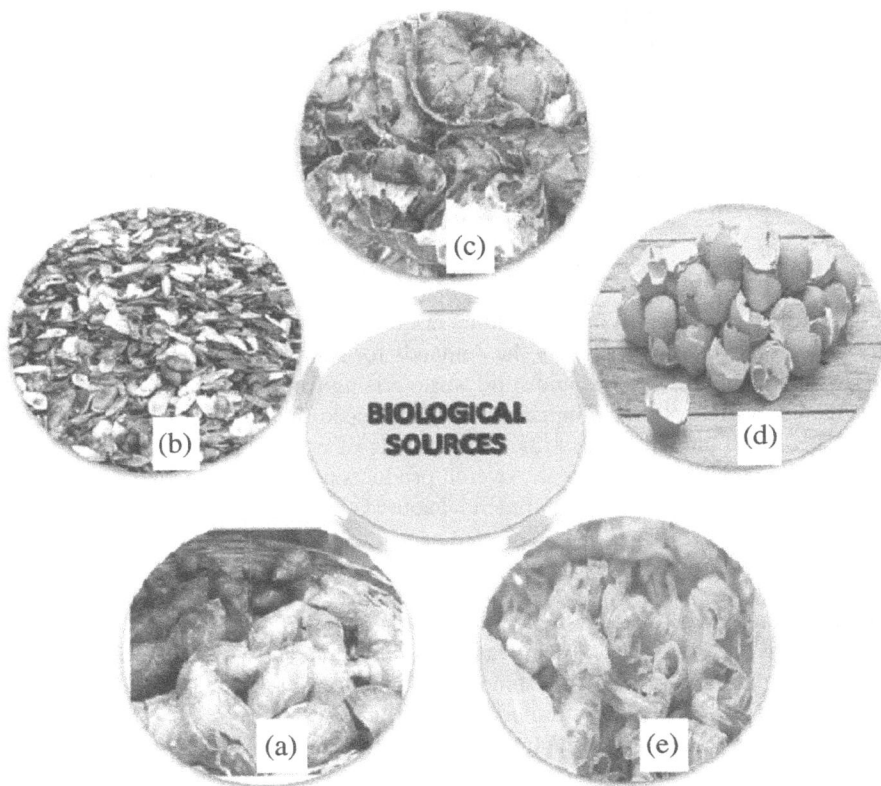

Figure 9.1 Calcium-based waste material from biological sources (a) Snail shells. (b) Mussel shell. (c) Crab shells. (d) Egg shell. (e) Shrimp shell.

Several environmental problems can be resolved with the help of biodiesel made from food waste like anthropogenic carbon emission resulting from the burning of fossil fuels and waste management [7]. The calcium rich waste materials are produced from waste materials of industries (like blast furnace slag, lime mud, carbide slag, and dolomite rock, etc.) and some biological sources (waste shells like egg shell, mollusk shells, etc) and animal bones. These wastes can also be employed as catalysts, which can both partially alleviate the disposal problem and lower the cost of synthesis [8]. The catalyst may be homogenous and heterogenous catalyst. The homogeneous frequently causes a variety of issues including environmental issue from the deposit of waste catalyst and equipment problems and their separation is very tedious process [9]. The use of heterogeneous catalyst could undoubtedly lower the price of biodiesel since wastage can be minimized, the reusability and environmentally friendly characteristics of heterogeneous catalyst make it easier to produce product and improve the product purity [10]. These waste shells can be used directly and the use through a range of customized techniques, e.g., loading of active compounds on to the shell, co-solvent method, organic solvent modification method, and hydration method. The advantage of calcium oxide (CaO) catalyst, like gentle reaction condition, non-poisonous, high basicity, inexpensive, economical, and no environmental impact have attracted a lot of interest [11]. The aim of current article is to give a variety of calcium-based waste products for catalysis that have been classified by their sources.

9.2   BIOLOGICAL SOURCES DERIVED CATALYST

It is possible to efficiently use a significant quantity of the waste produced by living things on every day to satisfy the demands for energy and other materials. This waste may be liquid, gaseous and solid waste arising from different sources. Few of the main contributors of biological wastes are: waste shells of eggs, mussel shells, snail shells and animal bones [12]. Waste shells of no values are used in place of commercial catalyst. To reduce the overall products costs by neglecting the requirement to obtained $CaCO_3$ from commercial sources is the use of waste shells [13].

9.2.1   EGGS SHELLS

The most common bird egg shell was chicken egg shell and eggshell production reached four million tonnes per year in China alone [14]. In the form of calcium carbonate and with a little amount of calcium phosphate [96% of $CaCO_3$, 2% Ca $(PO_4)_2$, 2% $MgCO_3$], organic material and water, eggshell is a significant natural calcium feedstock [15]. In numerous other processes like creation of hydrogen/syngas, synthesis of bioactive compounds, synthesis of DMC, synthesis of biodiesel, eggshell waste can be employed as a catalyst [16]. The transesterification reaction for biodiesel production is shown in Equation 9.1

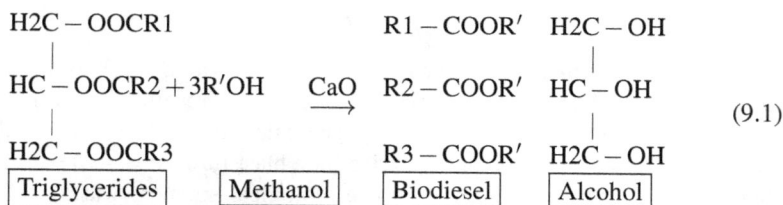

$$\begin{array}{l} H2C-OOCR1 \\ | \\ HC-OOCR2 +3R'OH \quad \xrightarrow{\text{CaO}} \\ | \\ H2C-OOCR3 \\ \boxed{\text{Triglycerides}} \quad \boxed{\text{Methanol}} \end{array} \quad \begin{array}{l} R1-COOR' \; H2C-OH \\ | \\ R2-COOR' \; HC-OH \\ | \\ R3-COOR' \; H2C-OH \\ \boxed{\text{Biodiesel}} \quad \boxed{\text{Alcohol}} \end{array} \quad (9.1)$$

Wei et al. [17] firstly prepared calcium oxide catalyst from chicken eggshell for transesterification reaction. Calcining egg shell at 1000 °C, CaO catalyst was easily obtained. The synthesize catalyst shows great performance and over 95% of FAME yield was produced under optimal parametric conditions.

Viriya et al. [18] synthesize CaO catalyst by calcined egg shell at optimal temperature of 800°C for 2–3 hours. The author found that 94.1% FAME yield (fatty acid methyl ester) was produced through palm olein oil.

9.2.2 MOLLUSK SHELLS

It has been demonstrated that CaO-based catalysts made from calcined mollusk shell can be employed effectively in a variety of applications [19]. Mollusk shells included snail shell, crab shell, mussel shell, clamshell, abalone, and shrimp shells, etc.

9.2.2.1 Snail Shells

They are mostly found in ditches and deep water. Apart from being used as jewellery and aesthetic thing, these shells don't really mean much. As a results shell is frequently used as a catalyst. Snail shell composed of calcium carbonate as main calcium components. The high temperature ranges more than $700°C$ is needed to convert calcium carbonate to calcium oxide which is used as a catalyst [12]. Birla et al. [20] prepared calcium oxide catalyst from calcined snail shell for biodiesel production. CaO catalyst obtained after calcining snail shells at $900°C$ and the prepared catalysts shows great catalytic performance in transesterification of waste frying oil. At optimal reaction condition, 87.28% of total biodiesel was obtained.

9.2.2.2 Oyster Shells

Oyster shells is the waste produced from shellfish farm and large amount of oyster usage resulted in large quantity of waste oyster shell [12]. Furthermore, oyster shell is a promising material to acquire calcium through preliminary treatment. Nakatani et al. [22] prepared CaO catalyst from oyster shell and investigated that when oyster shell heated at temperature greater than $700°C$ for 2 h shows great catalytic activity for transesterification of soybean oil and 73.8% of biodiesel was produced with high purity of biodiesel (98.4 wt%).

Lee et al. [23] synthesize CaO catalyst from seasnail shell for transesterification to produce biodiesel. They found that after calcination temperature at $800°C$ of shell, the percentage amount of calcium oxide was more than 98%. Molar ratio of (OH: oil) 12:1 and 5 wt% of catalyst loading, 86.75% biodiesel was produced in 6 h.

9.2.3 USE OF MODIFIED MOLLUSK SHELLS

For the preparation of highly efficient and good performer CaO catalyst number of modified techniques have been explored for eg. Co-solvent method, loading of active compounds and hydration method etc. [19]. To enhance the catalytic activity potassium salts (KBr, KI, KF) and barium salts have been demonstrated to be a type of great catalytic active phase. These salts could be impregnated on CaO Catalyst shells.

Jairam et al. [24] prepared a highly efficient catalysts by impregnated KI on oyster shells. Researchers revealed that, the surface area was enhanced through potassium iodide contribution on catalyst surface and biodiesel yield was also increased. In another study, Boro et al. [25] synthesize CaO catalyst from natural shell of Turbonilla striatula and utilize it in the transesterification of waste cooking oil (WCO).

The basicity of catalyst highly affects the catalytic activity of catalyst and biodiesel production was up to 98% under optimum reaction condition.

A co-solvent method is another technique to enhance the rate of reaction and improving the quality of biodiesel production [26]. Acetone, diethyl ether, tetrahydrofuran, etc. are used as co solvents. Roschat et al. [27], using tetrahydrofuran as a co solvent, prepared highly efficient catalyst and a precursor used was river snail shell for biodiesel production. Through this method, 98.5% of biodiesel was produced.

A hydration method is recently well focused by researcher in their efforts toward modification of highly efficient catalyst. Chen et al. [28], found that with the addition of organic solvents, the surface area and basicity was increased and crystalline size of catalyst at 100°C was decrease, that ascribe to enhance the catalytic performance of catalyst. The higher production yield of FAME with EtOH modification could reach up to 96.2%.

9.2.4 MODIFIED USE OF EGG SHELLS

The number of techniques were utilized to increase the reaction rate of catalysis processes catalysed by egg shell derived catalyst, e.g., loading of active compounds and ATT (assisted transesterification technique) [29].

Boro et al. [30], synthesize egg shell derived CaO catalyst by loading of active compound lithium salt. A CaO catalyst made of chicken egg shell doped with lithium was produced for the production of biodiesel through unusable feedstock (Nahor oil). Catalyst exhibited high activity with the loading of 2 wt% Lithium and 94% conversion of biodiesel was obtained under optimal reaction condition. Loading of active compound on porous support is a common way to obtained the desired physical properties and high catalytic activity of solid catalyst.

Chao and Seo [31], prepared CaO catalyst from egg shell through acid treatment method for biodiesel production (Figure 9.2). They were found that after acid treatment with HCl solution of quail egg shell and calcined at more than 800°C. (The quail eggs are the tiniest of all the bird eggs, measuring about half as large as a chicken egg [32]). The biodiesel conversion was up to 98.5% with resulting catalyst during repeated fivefold usage at 65°C and methanol to oil molar ratio 12:1 with 1.5 wt% catalyst. Quail egg shell's high catalytic activity was due to its tightly formed big pores which could offer a lot of strong basic sites and effective route for quick diffusion of oil molecules.

To obtaining the highest yield, ATT method is used to overcome the need of reactor operation conditions. There are two types of ATT's are most commonly used for the production of biodiesel from oil and animal fat (microwave assisted and ultrasonic assisted method) [33].

Microwave radiation is a form of electromagnetic energy that can cause intense and concentrated heating by directly transferring energy to reactant. As a result, it is possible to shorten the reaction time by eliminating the preheating step. Ultrasound is a form of assisted transesterification, involves creating a pressure wave with high frequency that cannot be heard by human [33]. Rather fine emulsion can be obtained

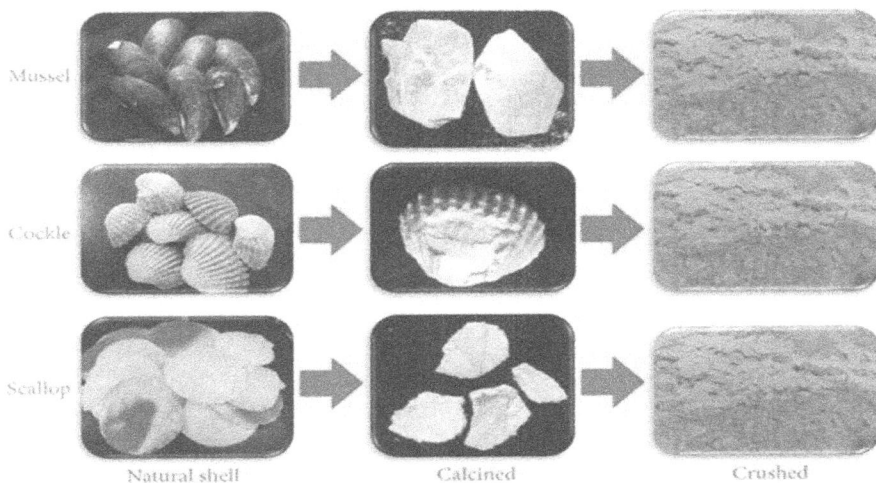

Figure 9.2   Treated waste natural shells at high temperature to obtain CaO catalyst.

after passing ultrasonic wave from the reaction mixture of vegetable oil and alcohol. These emulsions provide more reaction sites because of large surface area [34].

Khemthong et al. [18], applied microwave assisted technique to find out catalytic performance of CaO Catalyst prepared through calcined chicken egg shells for biodiesel production. The maximum yield of FAME was about 96.7% with microwave heating of 900 W. In another study, Chen et al. [35], synthesize CaO catalyst derived from ostrich egg shell by the use of ultrasonic assisted technique. Under optimal reaction condition the biodiesel yield obtained up to 92.7%.

Comparison of catalytic performance of CaO catalyst prepared from different sources

The waste ostrich egg shell and chicken egg shell were compared for the transesterification reaction in which they used as precursor. The CaO catalyst derived from calcined ostrich egg shell was found to be more active because of its high interfacial area, higher basicity and smaller crystalline size as compared to calcined chicken egg shell. Under optimum reaction condition 96% and 94% biodiesel yield was obtained through ostrich egg shell and chicken egg shells, respectively [32].

Correla et al. [36], compared the catalytic activity of crab shell and egg shell derived CaO catalyst in transesterification of vegetable oil and alcohol. They found that calcined egg shell exhibited large surface area and high basic sites and more active as compared to crab shell.

The catalytic property of calcium oxide catalyst derived from eggshell, golden apple snail and meretrix venus shell were elucidated to be active in heterogenous transesterification for biodiesel production. These all shells were calcined at same temperature and time of 800°C for 4 hour. Their catalytic activity was compared with different reaction time. At a reaction time of 1 hour, %FAME yield was in the order of eggshells (93%), golden apple snail (85%), and meterix venus shell (74%). On the other hand, with the reaction time of 2 hour the %FAME yield of eggshell (94%), golden apple snail (93%) and meretrix venus shell (94%). At a reaction time of 2 hour, greater than 90% of FAME yield was obtain by these catalysts. The catalyst made from egg shell has the biggest surface area and smallest particle size among the three raw materials which can lead to enhance the FAME yield percentage [37].

9.2.5   INFLUENCE OF MOLAR RATIO

The influence of methanol to oil ratio on FAME yield produced in transesterification through CaO catalyst synthesize from meretrix venus shell calcined at 800°C for 4 hour was shown in Figure 9.3(a), with optimal reaction parameters, 10 wt% of catalyst loading with 60°C temperature with various methanol to oil ratio. It demonstrates that by increasing methanol to oil ratio from 9 to 12, the biodiesel yield was enhanced significantly.

A greater concentration of methanol would encourage the development of methoxy species on the surface of CaO and should also cause the reaction equilibrium to shift in the direction of increased FAME yield. Additionally, methanol to oil ratio increases up to 18 were ineffective in promoting the reaction. It was thought that the high methanol content helped to favourably activate the reversible reaction of transesterification of high biodiesel synthesis. Due to its reversible nature, it is probable that, the transesterification which is reversible in nature occur between glycerol and FAME product. And they formed a product (i.e. monoglycerides and diglycerides) that can homogenize the solution. It may cause biodiesel production to decrease after increasing concentration up to 18 [37].

9.2.6   INFLUENCE OF CALCINATION TEMPERATURE AND TIME

Figure 9.3(b) demonstrated that the catalyst derived from venus shell was used to find out the influence of calcination temperature on biodiesel production. When shell is calcined at 700°C and 800°C for 4h with reaction conditions 10 wt% of catalyst amount, 60°C temperature with 15:1 molar ratio, the FAME yield was reached up to 37.8% and 87.4%, respectively. On the other hand, when calcination temperature increased to 900 and 1000°C, the FAME yield was decreased around 40.2% and 37.9%.

The influence of calcination time on biodiesel production was shown in Figure 9.3(c): When shell was calcined at 800°C with various heating time (0.5, 2, 4, 8 hours) and reaction conditions was 12:1 methanol to oil ratio, catalyst loading of 10 wt% with 60°C temperature. It was found that when shell was calcined for 0.5 hour, the obtained catalyst would not enough for better FAME yield and can't show higher activity. On contrary, the sample was calcined for 2–4 hours, attained higher activity for transesterification reaction. However, by increasing calcination time for 4 to 8 hours the biodiesel production was decreased dramatically [37].

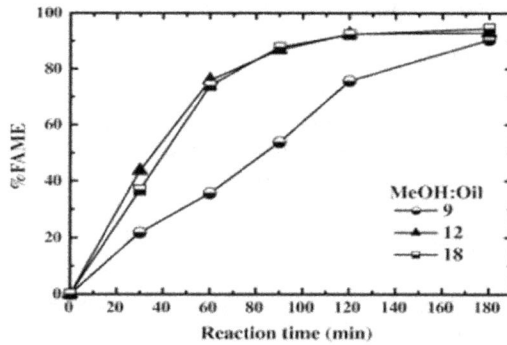

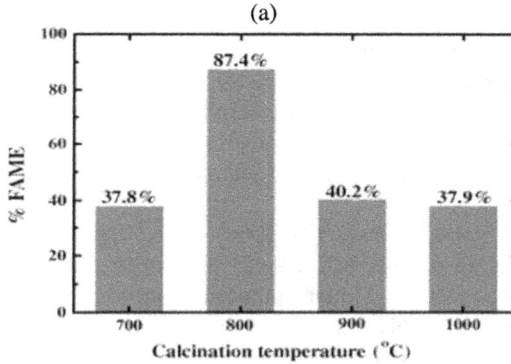

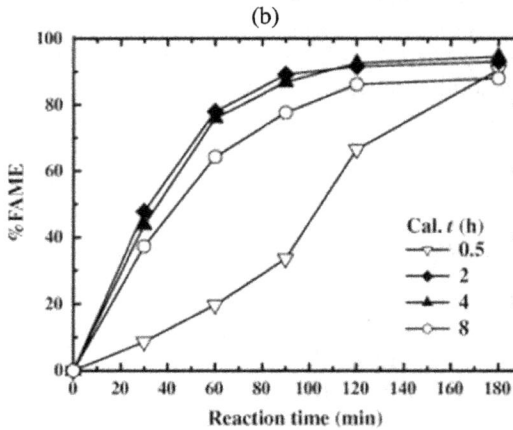

Figure 9.3 (a) Influence of reaction time on FAME yield with various methanol to oil molar ratios. (b) Effect of various calcination temperature on FAME yield over meretrix Venus shell calcined for 4 hours. (c) Effect of reaction time dependence on Biodiesel production during transesterification through shell calcined at 800°C with various calcination time.

9.2.7 ANIMAL BONES DERIVED CATALYSTS

Waste animal bone is another kind of Calcium based waste material which is widely available in nature. It comprises of $[Ca_{10}(PO_4)_6(OH)_2HAP]$ as its main components [38]. After the bone was calcined at high temperature the $Ca_3(PO_4)_2$ and calcium oxide phases could be formed and used as the active components in transesterification reaction. Waste bones are also used as in calcined form and with some modification methods.

Madhu et al. [39], synthesize CaO catalyst by utilizing the discarded part of fish for transesterification reaction. Under experimental condition, 96% of biodiesel conversion was obtained. Farooq et al. [40] utilize chicken bone for catalysis of transesterification reaction which contained similar component with domestic animal bones. The waste chicken bone achieved great performance for transesterification of low FFA contained waste cooking oil (WCO). After calcined at 900°C , a highly dense active basic sites produce on its surface and 89.33% yield of biodiesel was obtained.

The low surface area of waste bones derived catalyst is a perennial issue that affect biodiesel yield. To address this issue, researcher have led to modification on calcined animal bone by hydrothermal treatment [41] and impregnated in to solid support [42].

Using wet impregnation method, Chakborty et al. [43] prepared Ni/Ca/HAP (hydroxyapatite) and calcined waste fish scale as the support. The maximum conversion of FFA to FAME was 59.90%. Under optimum condition of 0.80 ml/min methanol flow rate, 30 wt% Ni $(NO_3)_26H_2O$ dosage and 300°C calcination temperature. To determine the optimal parametric values RSM (response surface methodology) method was used for esterification reaction. In the second step transesterification, using calcined fish scale base catalyst 98.40% yield of biodiesel was obtained. In another study, cost effective solid base catalyst was synthesize using calcined pig bone derived HAP as support for K_2CO_3 for biodiesel production. The resulting catalyst would achieve high basicity contributed to high catalytic activity and up to 96.4% biodiesel yield was produced under optimal reaction condition [44].

9.2.8 INDUSTRIAL WASTE DERIVED CALCIUM BASED CATALYST

Various mills, industries and factories worldwide generate a large amount of superfluous waste every year. Industrial waste (such as lime mud, carbide slag, blast furnace, etc.) are abundant worldwide (Figure 9.4). Therefore, those waste can also be used in terms of catalyst as shown in Table 9.1, which cannot only partly alleviated disposal problem but also reduce the synthesis cost of catalyst [12].

9.2.9 LIME MUD

Lime mud generates in pulp mills and paper making industry as by-product. The $CaCO_3$, unslaked $CaCO_3$ and Ca $(OH)_2$ are the main composition of lime mud [45]. $CaCO_3$ converted into calcium oxide by calcination at high temperature [50].

Hui et al. [43], prepared CaO catalyst by utilizing lime mud. They found the influence of calcination temperature of catalyst on transesterification reaction of peanut

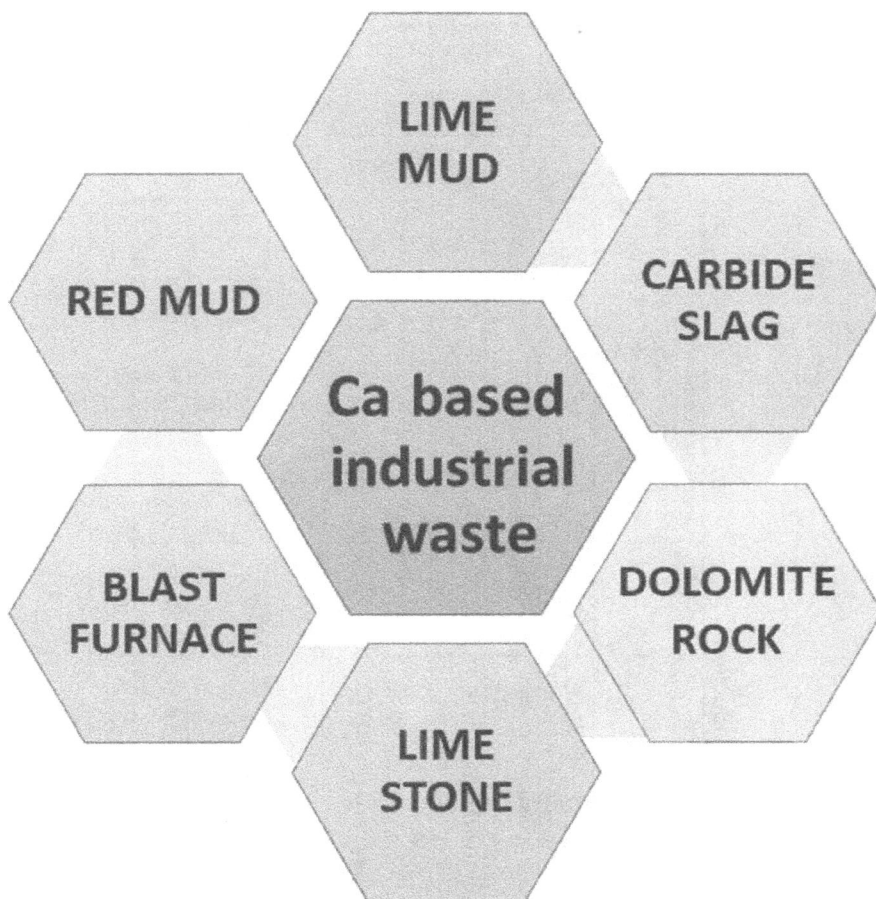

Figure 9.4 Calcium-based industrial waste materials.

oil with methanol. When calcination at 500–600°C temperature, the catalytic activity was negligible. But after calcination up to 800°C, the 90.51% of transesterification conversion was obtained. Catalytic reusabily of lime mud when tested for five cycle shows successful results.

Lie et al. [47] synthesize CaO catalyst from lime mud for the transesterification reaction. Researchers revealed that after impregnation with 20 wt% of KF and calcined at 600°C, the catalyst exhibited high catalytic activity due to its increased basic sites. Under optimal condition, 99.09% conversion of biodiesel was obtained.

9.2.10 CONSTRUCTIONAL LIME [LIME STONE]

Crashed limestone obtained from construction sites, it may be used as a precursor that can transform into active catalyst. This inexpensive transesterification of

Table 9.1

Different Type of Shells Derived Catalyst for Biodiesel Production

Raw Material	Feedstock Oil	Catalyst Preparation	Catalyst	Operating Conditions				Yield (%)	Reference
				Temp (°C)	Time (h)	Catalyst Amount (wt%)	Molar ratio (OH:OIL)		
Chicken eggshell	Soybean oil	Calcined at 1000°C for 2 hours	CaO	65	3	3	9:1	95	[17]
Golden apple Snail shell	Palm oil	Shell is calcined at 800°C for 4 hours	CaO	60	2	10	15:1	93.20	[58]
Chicken eggshell	Palm oil	Calcined for 4 hours at 800°C after adding 0.4 M Na$_2$SiO$_3$	CaO-SiO$_2$	65	8	9	15:1	80.2	[57]
Quail eggshell	Palm oil	Treated with 0.005M HCl, shell is calcined at 800°C for 2 hours	CaO	65	2	1	12:1	98	[31]
Oyster shells	Soybean oil	Calcined at 700°C for 3 hours	CaO	65	5	25	6:1	98.40	[22]
Abalone shell	Palm oil	Shell treated with ethanol at temp of 100°C and then calcined at 800°C for 4 hours	CaO	65	2.5	7	9:1	96.2	[28]
Fish bone	Soybean oil	Calcination temp of 997.42°C for 2 hours	Ca$_3$(PO$_4$)$_2$	70	5	1.01	6.27:1	97.73	[59]
Waste fish scale	Frying soybean oil	Calcined at 300°C for 4 hours through imperagenation with 30 wt% Cu(NO$_3$)$_2$.6H$_2$O	Ni/Ca/HAP	60	2	20	12:1	70	[38]
Sheep bone	Indian mustard oil	Sheep bone was calcined for 4 hours with temp of 909.4°C	Ca$_3$(PO$_4$)$_2$	70	3	4.97	10:1	91.22	[41]
Chicken bone	WCO with low FFA	Calcined at 900°C for 4 hours	Ca$_3$(PO$_4$)$_2$	65	4 h	5	15:1	89.33	[43]

vegetable oil with minor treatment [48]. Kozu et al. [49], prepared CaO catalyst from crushed lime stone. They found that at calcination of 900°C, the crushed limestone was activated with an image size range 1–1.7 um. With catalyst particle agglomeration, the initial run saw a yield of roughly 60% FAME in 2h. By increasing methanol to oil ratio, 96.5% of yield was increased with reaction time of 2 h. Catalytic reusability of crushed limestone when tested for ten cycle, shows successful results.

9.2.11 SLAG

Slag is the byproduct of the separation of metals from their respective ores. Typically, it is a mixture of several metals and metal oxides, sulfides, and silicon dioxides [50].

9.2.12 CARBIDE SLAG

The solid by product of the hydrolysis process of calcium carbide in the industrial generation of ethyne gas is known as carbide slag. 71.09 % of CaO was presents in carbide slag [51]. Li et al. [52] calcined the carbide slag waste at 650°C for successful synthesis of CaO catalyst for biodiesel production and resulting catalyst exhibited high basicity which contributed to its high catalytic activity. At 65°C temperature and in 30 min with 9:1 methanol to oil ratio with 1 wt% of catalyst, total 91.3% of FAME yield was achieved.

9.2.13 BLAST FURNACE

Blast furnace slag is another kind of industrial waste originating from iron manufacturing and processing. It contains various silicates, alumino, silicates, and metals [53]. Kuwahra et al. [54] synthesise hydrocalumite from blast furnace slag and hydrocalumite changes to CaO, MgO and other ternary oxides by calcination temperature of 800°C . Methanol to oil ratio 12:1,1 wt% of catalysts amount with 60°C temperature for 6 hours, 97% 0f FAME yield was produced.

9.2.14 DOLOMITE ROCK

Dolomite rock is proved to be a natural source of calcium consist of $MgCO_3$ and $CaCO_3$ in large amount. It possess high basicity and cost effective [57]. Ngamcharussrivichai et al. [55], carried out experiment with natural occurring dolomite rock calcined at 800°C used as a CaO catalyst. Under the optimal reaction condition, 99.9% FAME yield was produced with methanol to oil ratio 15:1, 10 wt% of dolomite catalyst calcined at 800°C with reaction time of 3 hours. In another study, Ilgem et al. [56] calcined dolomite at 850°C for the activation of catalyst for the conversion of canola oil. They found that, 91.78% yield was obtained under optimal condition of (methanol/oil) molar ratio 6:1, 3 wt% of catalyst loading, with the reaction temperature of 67.5°C for 3 hour. 90% yield was further produced after recycle the catalyst for 3 times.

Table 9.2

Different Type of Industrial Waste Derived CaO Catalyst for Biodiesel Production

Raw Material	Feedstock Oil	Catalyst Preparation	Operating Conditions						
			Catalyst	Reaction Temp. (°C)	Reaction Time (h)	Catalyst Amount (wt%)	Molar Ratio (OH:Oil)	Yield (%)	References
Lime stone	Soybean oil	Under the flow of helium gas, lime stone calcined at 900°C for 1.5 hours	CaO	64.7	1	/	12:1	93	[60]
Carbide slag	Soybean oil	Calcined carbide slag at 650°C	CaO	65	0.5	1.0	9:1	91.3	[49]
Lime mud	Refined peanut oil	Calcined lime at 800°C	CaO	64	3	6	15:1	94.40	[47]
Red mud	Soybean oil	Red mud is calcined at a temp of 200°C for 4 hours	CaO	65	3	4	24:1	94	[61]
Dolomite rock	Palm kernel oil	Rock is calcined at 700°C and then through Ca (NO₃)₂ Precipitation and calcined again at 800°C	CaO	60	3	10	15:1	99.90	[56]

9.3 FUTURE CHALLENGES AND PERSPECTIVE

CaO based catalysts could be employed as desirable and effective catalyst in catalytic processes according to the majority of study. As discussed in previous section, the calcium rich waste material derived catalyst only makes the process of biodiesel production more affordable and sustainable but also can counter the environmental complications. Below are some crucial issues that demands extensive research.

- Creation of catalyst with high selectivity, high renewability and minimal deactivation.
- Due to the reaction's leaching and saponification, calcium- based catalyst are susceptible to activity loss. Thus, more effective catalyst must be developed.
- Even though there is a surplus of waste materials that are rich in calcium, heterogenous catalysis is only feasible if their activity is maintained after repeated runs.

9.4 CONCLUSION

The use of waste material for catalysis not only serve to solve the problem related to waste disposal but also aids in the pursuit of a more sustainable and economical production of materials. The large scales use of those waste will offer an economic way to the utilization of calcium waste material and convert those waste into value added products. Synthetic routes to make effective catalyst can employ either a waste material alone or their combination with additional commercial component. This review has explored the series of calcium-based waste material from different sources like biological sources and industrial waste. These calcium rich waste material is utilized in many catalytic processes (Table 9.2). In present review, the CaO catalyst are synthesize from different sources and investigate their catalytic performance in transesterification reaction. Thereby the design of highly efficient and catalytically stable CaO catalyst which can be utilized in catalysis is still an important challenge. And a significant improvement is still necessary in the design of waste material catalyst to minimise the need of commercial sources such as the use of inedible feedstock and to reduce the cost of processing steps necessary to create the desired final catalyst.

REFERENCES

1. J. A. Bennet, K. Wilson, and A. F. Lee. Catalytic acivity of waste derived material. *Journal of Materials Chemistry A*, **4**:3617–3637, 2016.
2. R. Shan, J. Shi, B. Yan, G. Chen, J. Yau, and C. Liu. Transesterifications of palm oil to fatty acid methylester using K_2CO_3/polygarskite catalyst. *Energy Conversion and Management*, **116**:142–149, 2016.
3. A. F. Lee, J. A. Bennet, J. C. Manayil, and K. Wilson. Heterogenous catalysis for sustainable biodiesel production via esterification and transesterification. *Chemical Society Reviews*, **43**:7887–7916, 2014.

4. J. Shi, R. Shan, B. Yan, G. Chen, J. Yau, and C. Liu. Transesterification of palm oil to fatty acid methyl ester. *Energy Conversion and Management*, **116**:142–149, 2016.

5. I. Reyero, G. Arzamendi, and L. M. Gandia. Heterogenization of the biodiesel synthesis catalysis: CaO and novel calcium compound as transesterification catalyst. *Chemical Engineering Research and Design*, **92**:1519–1530, 2014 .

6. H. V. Lee, Y. H. Yap-Taufiq, M. Z. Hussain, and R. Yunus. Transesterification of jatropha oil with methanol over Mg-Zn mixed metal oxide catalysts. *Energy*, **49**:12–18, 2013.

7. M. Catario, M. Ramos, A. P. Soares Dias, M. T. Santos, J. F. Puna, and J. F. Gomes. Calcium rich food wastes based catalysts for biodiesel production, *Waste & Biomass Valorization*, **8**:1699–1707, 2017.

8. J. S. J. Ling, Y. H. Tan, N. M. Mubark, J. Kansedo, A. Saptoro, and C. N. Hipolito. A review of heterogenous calcium oxide based catalyst from waste for biodiesel synthesis. *SN Applied Sciences*, **1**:810, 2019.

9. G. Arzamendi, E. Arguinarena, I. Campo, S. Zabala, and L. M. Gandia. Alkaline and alkaline earth metal compounds as catalysts for the methanolysis of sunflower oil *Catalysis Today*, **133**:305–313, 2008.

10. C. R. V. Reddy, R. Oshal, and J. G. Verkade, Room temperature conversion of soybean oil and poultry fat to biodiesel catalysed by nanocrystalline calcium oxide. *Energy Fuel*, **20**:1310–1314, 2006.

11. M. Catario, M. Ramos, A. P. Soares Dias, M. T. Santos, J. F. Puna, and J. F. Gomes. Calcium rich food wastes-based catalyst for biodiesel production. *Waste & Biomass Valorization*, **8**:1699-1707, 2017.

12. A. Marwaha, P. Rosha, S. K. Mohaputra, S. K. Mohla, and A. Dhir. Waste material as a potential catalysts for biodiesel production: Current state and future scope. *Fuel Processing Technology*, **181**:175–186, 2018.

13. D. A. Oliveira, P. Benelli, and E. R. Amante. A literature review on adding values to solid residues: Egg shells. *Journal of Cleaner Production*, **46**:42–47, 2013.

14. M. Balakarishanan, V. S. Batra, J. S. J. Hargreaves, and I. D. Pulford. Wastes material catalytic opportunity: An overview of the application of largescale waste material as resources for catalytic application. *Green Chemistry*, **13**:16–24, 2011.

15. A. Laca, L. Adriayan, and D. Mario. Egg shell waste as a catalyst: A review. *Journal of Environmental Management*, **197**;351–359, 2017.

16. A. Hart and E. Aliu. Materials from eggshells and animal bones and their catalytic applications. *Design and Application of HPA Based Catalysts*. 437–479, 2022.

17. Z. Wei, C. Xu, and B. Li, Applications of waste eggshells as low-cost solid catalyst for biodiesel production. *Bioresource Technology*, **100**:2883–2885, 2009.

18. P. Khemthong, C. Luadthong, W. Nualpaeng, W. Changsuwan, P. Tongprem, and N. Viriya-Empikul. Industrial eggshell wastes as the heterogenous catalysts for microwave assisted biodiesel production. *Catalysis Today*, **190**:112–116, 2012.

19. R. Shan, C. Zhao, P. Lv, H. Yuan, and J. Yao. Catalytic application of calcium rich waste material for biodiesel: Current state and perspective. *Energy Conversion and Management*, **127**:273–278, 2016.

20. A. Birla, B. Singh, S. N. Upadhyay, and Y. C. Sharma. Kinetics study of synthesis of biodiesel from waste frying oil using heterogenous catalyst derived from snail shell. *Bioresource Technology*, **106**:95–100, 2012.

21. Y. C. Lin, K. T. T. Ameesho, C. E. Chen, P. C. Cheng, and F. C. Chou. A cleaner process for green biodiesel synthesis from waste cooking oil using recycled waste oyster shell as

a sustainable base heterogenous catalyst under microwave heating system. *Sustainable Chemistry and Pharmacy*, **17**:100310, 2020.

22. N. Nakatani, H. Takomori, K. Takeda, and H. Sakagawa. Transesterification of soybean oil using combusted oyster shell waste as a catalyst. *Bioresource Technology*, **100**:1510–1513, 2009.

23. S. L. Lee, Y. C. Wong, Y. P. Tan, and S. Y. Yew. Transesterification of palm oil to biodiesel using waste obtuse horn shell derived CaO catalyst. *Energy Conversion and Management*, **93**:282–288, 2015.

24. S. Jairam, P. Kolar, R. S. Sharma, J. A. Osburne, and J. P. Devis, KI Imperagenated on oyster shell as solid catalysts for soybean oil transesterification. *Bioresource Technology*, **104**:329–335, 2012.

25. J. Boro, L. J. Konwar, A. J. Thakur, and D. Deka. Ba-doped calcium oxide catalyst derived from waste shells of T. striatula as heterogenous catalyst for biodiesel production. *Fuel Processing Technology*, **92**:2001–2007, 2011.

26. Y. Alhassan, N. Kumar, I. M. Bugaje, H. S. Pali, and P. Kathkey. Co-solvent transesterification of cotton seed oil into biodiesel: Effect of reaction condition on quality of fatty acid methyl ester. *Energy Conversion and Management*, **84**:640–648, 2014.

27. W. Roschat, T. Siritanon, T. Kaewpuang, B. Yoosuk, and V. Prowark. Economical and green biodiesel production processes using river snail shell derived heterogenous catalyst and co solvent method. *Bioresource Technology*, **209**:343–350, 2016.

28. G. Chen, R. Shan, B. Yan, J. Shi, S. Li, and C. Liu. Remarkably enhancing the biodiesel yield from palm oil upon abalone shell derived CaO catalyst treated by ethanol. *Fuel Processing Technology*, **143**:110–117, 2016.

29. P. L. Boey , G. P. Maniam, and S. A. Hamid. Performance of CaO as heterogenous catalyst in biodiesel production: A review. *Chemical Engineering Journal*, **168**:15–22, 2011.

30. J. Boro , L. J. Konwar, and D. Deka. Transesterification of nonedible feedstock with lithium incorporated egg shell derived CaO for biodiesel production. *Fuel Processing Technology*, **122**:72–78, 2014.

31. Y. B. Chao and G. Seo. High activity of acid treated quail egg shell catalyst in transesterification of palm oil with methanol. *Bioresource Technology*, **101**:8515–8519, 2010.

32. Y. H. Tan, M. O. Abdullah, and C. H. Nolasco. The potential of waste cooking oil based biodiesel using heterogenous catalyst derived from various calcined egg shells coupled with an emulsification technique: A review on the emission of reduction and engine performance. *Renewable and Sustainable Energy Reviews*, **45**:574–588, 2015.

33. P. Adiwale, J. M. Dumant, and M. Ngadi. Recent trends of biodiesel production from animal fat waste and associated production techniques. *Renewable and Sustainable Energy*, **47**:598–603, 2015.

34. A. S. Badday, A. Z. Abdullah, and K. T. Leu, Optimization of biodiesel production process from jatropha oil using supported heteropoly acid catalyst and assisted by ultrasonic energy. *Review Energy*, **50**:427–432, 2013.

35. G. Chen, R. Shan, J. Shi, and B. Yan, Ultrasonic assisted production of biodiesel from transesterification of palm oil with methanol over ostritch egg shell derived CaO catalysts. *Bioresource Technology*, **171**:428–432, 2014.

36. L. M. Corella, R. M. Saboya, S. N. Campelo, J. A. Cecila, E. C. Rodriguez, and C. L. Cavalcante Jr. Characterization of calcium oxide catalysts from natural sources and their application in the transesterification of sunflower oil. *Bioresource Technology*, **151**:207–213, 2014.

37. N. Viriya Empicul, P. Krosae, W. Nualpaeng, B. Yoosuk, and K. Faungnawake. Biodiesel production over Ca-based solid catalysts derived from industrial waste. *Journal of Fuel*, **92**:239–244, 2012.
38. R. Chakraborty, S. Das, and S. K. Bhattacharjee. Optimization of biodiesel production from Indian mustard oil by biological tri-calcium phosphate catalyst derived from turkey bone ash. *Clean Technologies and Environmental Policy*, **17**:455–463, 2015.
39. D. Madhu, B. Singh, and Y. C. Sharma. Studies on application of fish waste for synthesis of high-quality biodiesel. *RSC Advances*, **4**:31462–31468, 2014.
40. M. Farooq, A. Ramli, and A. Naeem. Biodiesel production from low FFA waste cooking oil using heterogenous catalyst derived from chicken bone. *Renew Energy*, **76**:362–368, 2015.
41. C. Chingakham, C. Tiwary, and V. Sajith. Waste animal bone as a novel layered heterogenous catalysts for the transesterification of biodiesel. *Catalysis Letter*, **149**:1100–1110, 2019.
42. V. Volli, M. K. Purkait, and C. M. Shu. Preparation and characterization of animal bone powder impregnated fly ash catalysts for transesterification. *Science of Environment*, **669**:314–321, 2019.
43. R. Chakraborty and D. Roy Chowdhury. Fish bone derived natural hydroxyapatite supported copper acid catalyst: Taguchi optimization of semi batch oleic acid esterification. *Chemical Engineering Journal*, **215**:491–499, 2013.
44. G. Chen, R. Shan, J. Shi, C. Liu, and B. Yan. Biodiesel production from palm oil using active and stable K doped HAP catalyst. *Energy Conversion and Management*, **98**:463–469, 2015.
45. J. Qin, C. Cui, A. Hussain, C. Yang, and S. Yang. Recycling of lime mud and fly ash for fabrication of anorthite ceramic at low sintering temperature. *Ceramics International*, **41**:5648–5655, 2015.
46. H. Liu, S. Niu, C. Liu, M. Liu, and M. Huo. Uses of lime mud from paper mill as a heterogenous catalysts for transesterification. *Science China Technological Sciences*, **57**:438–444, 2014.
47. H. Li, S. Niu, C. Lu, M. Liu, and M. Huo, Transesterification catalysed by industrial waste lime mud doped with KF and the kinetics calcination. *Energy Conversion and Management*, **86**:110–117, 2014.
48. R. Ghanei, G. Maradi, A. Heydarinasab, A. A. Seifkordi, and M. Ardjmand. Utilization of constructional lime as heterogenous catalyst in biodiesel production from waste frying oil. *International Journal of Environmental Science and Technology*, **10**:847–854, 2013.
49. M. Kouzu, J. S. Hidaka, Y. Komichi, H. Nakano, and M. Yamamoto. A process to transesterification vegetable oil with methanol in the presence of quick lime bit functioning as solid base catalyst. *Fuel*, **88**:1983–1990, 2009.
50. R. Sun, Y. Li, C. Liu, and C. Lu. CO2 capture using carbide slag modified by propionic acid in calcium looping process for hydrogen production. *International Journal of Hydrogen Energy*, **38**:13655–13663, 2013.
51. F. Li, H. Li, L. Wang, and Y. Cao, Waste carbide slag as a solid base catalyst for effective synthesis of biodiesel via transesterification of soybean oil with methanol. *Fuel Processing Technology*, **131**:421–429, 2015.
52. D. W. Lewis. Properties and usage of iron and steel slags. National Slag association presented at symposium on slag National Institute for transport and road research South Africa, 182–186, 1982.

53. Y. Kuwahara, K. Tsuji, T. Ohmichi, T. Kamegava, K. Mori, and H. Yamashita. Transesterification using a hydrocalumite synthesized from waste slag: An economic and ecological route for biofuel production. *Catalysis Science & Technology*, **2**:1842–1851, 2012.

54. S. Gunasekaran and G. Anbalagan. Thermal decomposition of natural dolomite. *Bulletin of Materials Science*, **30**:339–344, 2007.

55. C. Ngamcharussirivichai, W. Wiwantnimit, and S. Wangnoi. Modified dolomite as catalyst for palm kernel oil transesterification. *Journal of Molecular Catalysis A: Chemical*, **276**:24–33, 2007.

56. O. Ilgen. Dolomite as a heterogenous catalyst for transesterification of canola oil. *Fuel Processing Technology*, **92**:452–455, 2011.

57. G. Chen, R. Shan, S. Li, and J. Shi. A biomimetic silicification approach to synthesize CaOSiO2 catalyst for the transesterification of palm oil into biodiesel. *Fuel*, **153**:48–55, 2015.

58. N. Viriya-Empikul, P. Karasae, W. Nualpaeng, B. Yoosuk, and K. Faungnawakji. Biodiesel production over calcium based solid catalyst derived from industrial waste. *Fuel*, **92**:239–244, 2012.

59. R. Chakraborty, S. Bepari, and A. Banerjee. Application of calcined waste fish (Labeo rohita) scale as a low-cost heterogenous catalyst for biodiesel synthesis. *Bioresource Technology*, **102**:3610–3618, 2011.

60. M. Kouzu, T. Kasuno, M. Tejika, Y. Sugimoto, S. Yomanaka, and J. Hidaca. Calcium oxide as a solid base catalyst for transesterification of soybean oil and its application to biodiesel production. *Fuel*, **87**(12):2798, 2008.

61. Q. Liu, R. R. Xin, C. C. Li, C. L. Xu, and J. Yang. Applications of red mud as a basic catalyst for biodiesel production. *Journal of Environmental Sciences,* **57**(2):823, 2014.

10 Integration of IoT and AI in Bioengineering of Natural Materials

Jaswinder Singh Sidhu, Abhinav Jamwal, and Devinder Mehta
Department of Physics, Panjab University, Chandigarh, India

Aayush Gautam
Department of Electronics and Communication Engineering,
University Institute of Engineering and Technology
Panjab University, Chandigarh, India.

10.1 INTRODUCTION

Biomedical engineering or Bioengineering is a broad field that comprises a vast realm including development and designing of medical devices and detectors, tissue engineering, genetic engineering, prosthetics engineering, pharmaceutical engineering, and neural engineering. It is an amalgam of biotechnology and biomedical sciences aimed to contribute to the development of new technologies that can address various healthcare challenges, and advance medical knowledge. In this artificial intelligence-driven era, people are actively engaged in the integration of Artificial Intelligence (AI) and Internet of Things (IoT) with biomedical engineering [1]. AI trains machines so that they can perform tasks akin to or better than human beings. Machine Learning (ML) is a subset of AI that focuses on training machines in such a way that the machines can execute tasks without any explicit instructions, only by identifying and learning from the patterns existing in the data. Deep Learning (DL) is the subset of machine learning that simulates human neural network to enable machines to perform complex tasks by identifying more complicated patterns [2]. Artificial intelligence and machine learning, due to its high accuracy and faster speed, is already being used in drug discovery, design and development [3], drug delivery [4], diagnosis [5], surgery [6], etc.

In this chapter, fundamentals of IoT, AI, ML, and DL (shown in Figure 10.1) and their applications in the field of bioengineering have been discussed. Further, this chapter emphasizes on the applications, advantages and challenges of using natural

DOI: 10.1201/9781003360599-10

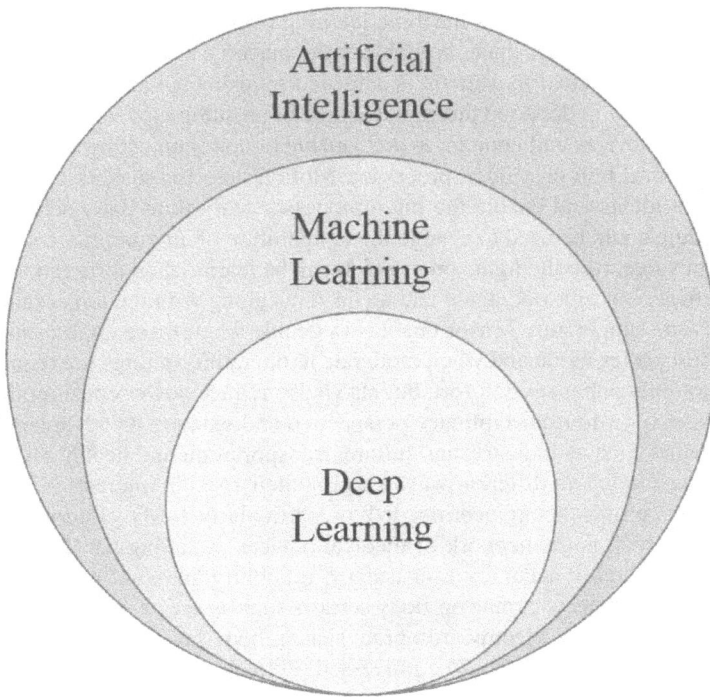

Figure 10.1 DL is a subset of ML which in turn is a subset of AI.

materials in bioengineering. This chapter mainly aims at the synergy of IoT and AI in the realm of bioengineering and using natural materials such as calcium phosphate with AI-driven technology.

10.2 IOT AND AI FUNDAMENTALS

10.2.1 IOT IN BIOENGINEERING

The interconnected network of devices (things) through which they can collect and exchange data is called Internet of Things (IoT). Internet word in IoT does not necessarily mean internet; data could be shared over the internet or cloud but this is not necessary. It can be done over local network also. Devices involved in IoT are generally equipped with transducers, actuators, controllers, and connectivity features enabling them to gather and share data with each other [7]. Transducers are basically sensors such as temperature sensor, proximity sensor, light detectors, etc. that convert variations in physical quantities such as temperature, magnetic field, light, etc., and convert it into corresponding electrical signal. This electrical signal is amplified, processed and converted into useful information which is then made available to other devices through any network so that the data collected could be invoked by any of the devices as and when required. Innumerable devices could be connected together and could use the data. This makes machine learning even easier when data from large number of environments is made available to the machine at the same time. Based on the input signal, physical movement is achieved by using actuators. The concept of IoT could be easily understood through example of an automated home. Suppose in a home, various devices and appliances are capable of communicating with each other through a central hub or a microprocessor. Motion detector sensors can detect the motion of residents and enable the microprocessor learn their daily schedule. This daily schedule could be used to change the temperature of air conditioners or heaters to optimum value, turn the lights on or off. It can be learnt when different appliances such as refrigerator are used again and again. This along with the temperature value acquired from temperature sensor enables to decide when these appliances need to be run in full power mode and when moderate temperature settings are required. IoT in home not only enhances comfort, but also helps reduce power consumption [8].

IoT possesses a multidisciplinary perspective and extends its advantages across various sectors such as industry, agriculture, transportation and health. Different researchers explain IoT in different ways keeping their specific interests in focus [9].

In healthcare and bioengineering, IoT or particularly IoMT (Internet of Medical Things), which is the network of medical devices including hardware and software can hep increase accuracy, consistency, reliability and swiftness in diagnosis and treatment. Various wearable on-body sensors such as accelerometer, temperature sensor, humidity sensor, electrocardiogram sensor, body position sensor, etc. can be used to collect data which could be processed, uploaded to cloud server and monitored in real-time for personal use or for sharing with healthcare service providers. In case any anomaly is there, help could be automatically called as per the requirement [10].

IoMT also allows Remote Patient Monitoring (RPM) (Figure 10.2) through which medical practitioners can keep a check on their patients health from a distance and can diagnose a problem before it becomes serious. IoMT also allows using diagnosis machines at Point of Care (PoC) and getting the data wirelessly [11, 12].

The integration of IoT with biomedical engineering has paved the way for ground breaking advancements. IoT could be used to pick up data directly from the brain and use it to control physical devices and instruments.

Various daily life activities such as eating, sleeping, etc. could be reflected by the Electroencephalography (EEG) data which could be collected by different types of EEG sensors [13] attached to the scalp. Laport et al. have developed a method to find relaxed state or state of high wakefulness and eye state using EEG [14]. Moreover, it is now possible to turn the appliances on or off, just by making a simple circuit making use of a microcontroller such as Arduino UNO, an EEG sensor and some other components and modules [15]. IoT has a lot to do in integration with biomedical engineering in the future.

10.2.2 AI IN BIOENGINEERING

Artificial intelligence (AI) is the idea, design, and development of computational systems and algorithms that can mimic cognitive capabilities associated with

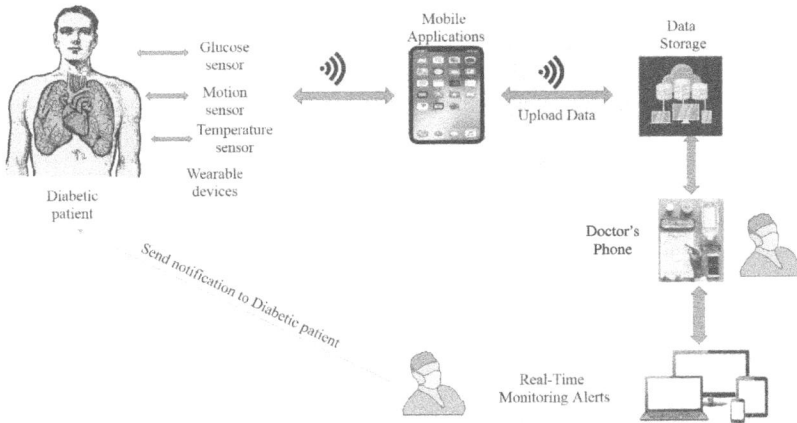

Figure 10.2 Remote Patient Monitoring (RPM).

human intelligence. These activities include tasks such as data learning, problem solving, adapting to changing conditions, and making decisions, which frequently rely on complex algorithms, neural networks, and machine learning approaches. AI is playing an increasingly important role in bioengineering, altering how we approach research, diagnostics, treatment, and overall healthcare. AI analyzes huge volumes of biological and clinical data using advanced algorithms and machine learning approaches, resulting in more accurate predictions, insights, and personalized solutions. AI is being used in drug discovery, development and treatment of several diseases through ML assisted drug delivery [3, 16]. AI has proved to be successful in predicting drug properties such as binding affinities, toxicity, dipole moment, solubility, strength of bonds and other chemical properties [17]. This is discussed in more detail in Section 10.4.2.

The use of machine learning enables the sophisticated analysis of data derived from various diagnostic instruments, reducing errors and speeding up the detection process. The incorporation of machine learning techniques can improve all diagnostic methodologies covering digital image processing. Furthermore, the use of AI substantially simplifies massive data handling [18].

AI holds enormous promise for pushing biotechnology to new heights. Tang et al. have recently developed an AI powered decoder that could be used to translate thoughts into text in real-time [19]. Companies like Neuralink are also working on developing thought translators and brain-computer interface [20, 21]. Though this technology is in its nascent stages, it has the potential to transform the realm of communication, diagnostics, therapeutics and biomanufacturing.

10.2.3 SYNERGY OF IoT AND AI

When smart devices come together with super smart computers, it becomes a powerful combination, that can achieve unprecedented success in the field of bioengineering. Various transducers collect vast information and AI can quickly make sense of that information. This synergistic approach can help make remote patient monitoring [22] and early diagnosis of diseases [23] possible. Large data is generated during drug discovery experiments by various instruments connected together (IoT) in laboratories. AI speeds up the analysis of data which in turn helps accelerate the process of identification of proper drug.

As discussed earlier, AI can detect small anomalies in the data collected by on-body IoT sensors that may elude human observation. By ML and DL, devices can improve their performance without being programmed explicitly and can enhance healthcare outcomes. As IoT and AI continue to harmonize in this field, they have the potential to revolutionize healthcare by enabling more exact diagnoses, expediting drug development, and enhancing bioprocessing processes.

10.3 NATURAL MATERIALS IN BIOENGINEERING

10.3.1 SIGNIFICANCE OF NATURAL MATERIALS

Natural materials have a large number of biomedical applications due to their:

1. **High Biocompatibility:** Biocompatibility refers to the ability of a material to interact with biological systems without causing any undesirable effects. Derived from biological sources, natural materials have much more biocompatibility than synthetic materials. They have inherent resemblance to substances found in human body. Natural materials interact favorably with cells and tissues, promoting cell adhesion, proliferation, and migration. This trait is critical in regenerative medicine, where the goal is to assist tissue repair and replacement.
2. **Reduced Toxicity:** Natural materials do not carry any chemical additives, and therefore are much safer than synthetic materials to be used for medical implants, drug delivery and wound healing.
3. **Sustainability:** Most of the natural materials are renewable and sustainable. This makes them environment friendly and greener approach as they do not contribute to environmental pollution.
4. **Ease of Modification:** To meet specific requirements, natural material can be easily modified and functionalized through various physical and chemical processes. This makes them more versatile as their interaction with cells and tissues can be easily controlled.
5. **Diagnostic Applications:** Natural materials can be used as biosensors as they respond to specific conditions and can also be used to detect specific molecules.

10.3.2 CHALLENGES

1. **Consistency:** As natural materials are derived from biological sources, they can vary extremely in composition and properties. Same materials acquired from different sources may not be consistent with each other.

2. **Stability and Mechanical Strength:** Natural materials are not much stable as they degrade over time. It is challenging to develop techniques to improve the stability of these materials while retaining their desired properties. Some natural materials do not have required mechanical strength, and it is also challenging to increase their strength without altering their biocompatibility.

3. **Cost:** Some nature materials are pretty rare and require specialized cultivation. This affects the cost effectiveness of natural materials.

10.3.3 NATURAL MATERIALS AS IOT SENSORS

Calcium Phosphate materials are natural compounds having Ca^{2+} and PO_4^{3-} ions arranged in different configurations. There are four types of Calcium Phosphate- Hydroxyapatite, Brushite, Tricalcium Phosphate and Whitlockite. Calcium Phosphate materials are naturally found in bones and tooth enamel. Hydroxyapatite is the major component of bone material and is used in bone regeneration [24] and dental applications [25].

10.4 SYNERGY OF IOT AND AI IN BIOENGINEERING

10.4.1 IOT-DRIVEN DATA COLLECTION AND MONITORING

Bio engineering is one of the many industries that has been impacted by inclusion of IoT, reduction of man power, easier collection of data, better accuracy, innovative ways to approach problems being some of its major advantages. This in turn leads to advancement in research and technology. The world is moving towards technical advancements in all types the fields in order to increase the comfort of life. But at the same time, it has lead to sedentary lifestyle and higher rate of getting diseases. Thus, it is very important to use some sensors to identify hotspots of such problems so that efficient implementations could be made. The best way to achieve this is to integrate these sensors with in the devices that we use in everyday life. Smartwatches and smartphones are two of the most common devices in everyday use. Let us discuss these:

Smartwatches and Wearable Devices: Equipped with various sensors, smartwatches, and wearable devices may continuously monitor vital signs, physical activity, and sleep habits. They monitor heart rate [26], steps taken [27, 28], calories burnt, and even sleep quality. Individuals can obtain insights into their everyday activities and make informed lifestyle decisions by evaluating this data. Abnormalities in heart rate, sleep patterns, or abrupt changes in activity levels can serve as early signs of possible health concerns. Unusual heart rhythms observed by a smartwatch, for example, can urge users to seek medical assistance for illnesses **such as atrial fibrillation.**

Mobile Apps and Health Tracking: To aggregate and analyse health data, mobile apps can connect to wearable devices and other IoT sensors. They provide easy-to-use interfaces for tracking and visualizing health data. Wearable data can be

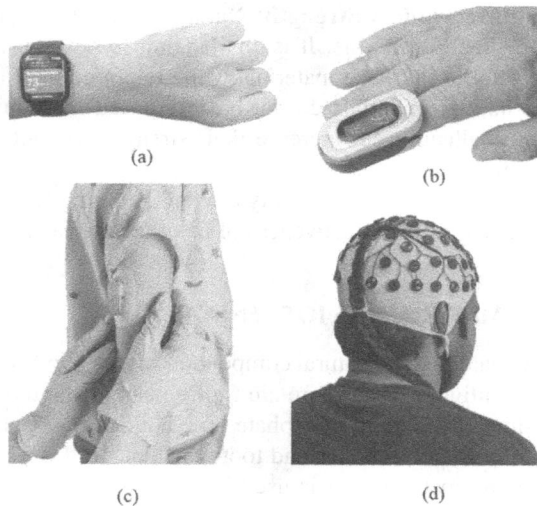

Figure 10.3 (a) Smartwatch having heart rate sensor. (b) Pulse Oximeter. (c) Continuous Glucose Monitor (CGM). (d) Electroencphalogram (EEG) cap.

processed by mobile apps to give consumers with personalized health insights and suggestions. These insights may include suggestions for bettering sleep quality, boosting physical exercise, or dealing with stress. Based on their health trends and development over time, users can make data-driven decisions. For example, if a users step count drops down drastically, the app may send motivating messages to encourage him/her to be more active (Figure 10.3). As the uses of sensors have been discussed, let us have a look in detail at what these sensors are capable to achieve. Following are some of the main sensors in use:

- **Heart Rate Monitors:** These sensors, which may be found in smartwatches and fitness trackers, measure the electrical activity of the heart. They give real-time heart rate data, which is useful for monitoring exercise intensity and heart health [26]. These monitors measure changes in heart rate using optical technologies or chest straps. Users can test cardiovascular fitness, track recuperation, and optimize workouts using heart rate data. Commonly used optical heart rate sensors generally produce accurate heart rate readings, irrespective of the age of user [29]. Some sensors also provide information on stress levels and general heart health, which helps with assessment of overall well-being. Heart rate sensors are being used to help diagnose abnormal heart rhythms and direct therapy choices.
- **Electrocardiogram (ECG) Sensors:** ECG sensors monitor the electrical activity of the heart over a period of time. These sensors record the heart's rhythm and waveform using electrodes that are applied to the skin. They are used to detect abnormal heart rhythms and provide information on cardiac

health [29]. ECG has always been a popular measurement scheme to assess and diagnose cardiovascular diseases(CVDs) [29]. By assisting in the diagnosis of arrhythmia, other heart diseases and associated hazards, ECG data sheds light on cardiac health. ECG sensors in wearable technology provide the ease of continuous heart monitoring, warning users of anomalies and raising general awareness of heart health. ECG sensors also aid in the early detection of heart problems, enhancing outcomes through prompt care.

- **Pulse Oximeters:** Pulse oximeters are sensors that monitor blood oxygen levels as well as pulse rate. They are frequently used to track respiratory problems. Ring-type wireless pulse oximeters have also been developed that can replace fingertip-type oximeters [30].

- **Accelerometers:** Accelerometers detect movement and acceleration. They are used in wearable devices to track steps, calculate distance, and determine activity levels throughout the day [28]. Accelerometers are essential for spotting falls in elderly people and help GPS-enabled devices navigate. Accelerometer data on smartwatches gave a high accuracy of over 91% with all five algorithms for walking and jogging activities [31].

- **Gyroscopes:** Gyroscopes measure orientation and rotation. Gyroscope sensors monitor angular velocity, making it possible to follow changes in direction and device orientation precisely. Smartphones, drones, and gaming controllers, all use gyroscopes to enable capabilities like screen rotation and gesture recognition, which improve user experience. Gyroscope sensors are necessary for electronic devices to stay balanced, stable, and responsive during navigation. Smartphone gyroscopes and those fitted in wearable devices can detect the body posture and can be used for health-related biofeedback applications in sports, recreation, rehabilitation, and well-being [32].

- **Continuous Glucose Monitors (CGMs):** Sensors for continuous glucose monitoring (CGM) allow diabetics to track their blood sugar levels in real time. These sensors examine the amounts of glucose in interstitial fluid and provide a constant stream of information to assist in controlling blood sugar. Alerts from CGM sensors for high and low glucose levels enable quick corrective action. Electrochemical-based continuous glucose monitoring devices have been commercialized and appreciated by patients [33]. They improve the convenience and quality of life for diabetic patients by reducing the need for regular finger-stick testing. Better glucose control provided by CGMs lowers the possibility of diabetes related complications.

- **Electromyography (EMG) Sensors:** EMG sensors detect electrical signals generated by muscle contractions. They record the patterns of muscle activity when attached to the skin or implanted in the muscles. EMG sensors are crucial instruments for comprehending how muscles move and operate. The data collected by such sensors is used to evaluate muscle performance, spot abnormalities, and create efficient rehabilitation plans. EMG sensors are also essential for creating improved prosthetics and exoskeletons that react to muscle signals and improve mobility for those with limb loss or mobility problems. The data could also be used to control appliances [34, 35].

- **Electroencphalogram (EEG) Sensors:** The electrical activity of the brain is measured using an electroencephalogram. The sensor, which captures brain waves, is applied to the scalp and it recognizes the minute electrical signals that the brain's neurons produce. The study of brain dynamics, sleep patterns, neurological illnesses, and medical diagnosis becomes easy with this information. It can also be used to the study the brain behavior for betterment of brain-computer interfaces that could be used to control external appliances also [13, 15].

Micro-electromechanical systems are gaining in popularity in the sensor industry now a days. MEMS use micro-engineering to integrate circuits and microscopic mechanical components into silicone microchips. In doing so, it is possible to create micro-scale sensors with a range of sensing capabilities.

Whilst some research focuses around use of MEMS sensors for specific healthcare applications, researchers are exploiting these technologies to create accessible sensor fusion health monitoring systems [36].

Figure 10.4 shows a force sensor next to the tip of a pencil. This really shows how much the world has advanced in terms of size of these sensors and how well they can be integrated in the circuits. Integration and examination of Health Data: Integrating data from numerous sources, such as smartwatches, mobile apps, and fitness equipment, provides for a comprehensive examination of an individual's health and lifestyle. Patterns of sedentary behaviour, sleep difficulties, or rapid changes in health indicators can be found by evaluating integrated data. These hotspots can point to regions that require action. Personalized therapies could be implemented with a better understanding of user behaviour and health patterns. Users can receive personalized recommendations, such as taking breaks while prolonged periods of sitting in the same posture or planning the workout schedule.

Figure 10.4 Size of force sensor compared to a pencil tip.

10.4.2 AI-POWERED ANALYSIS AND DECISION-MAKING

The use of advanced AI tools and algorithms to process, understand, and draw meaningful insights from complex and diverse data is referred to as AI-powered analysis and decision-making. These game-changing technologies help bioengineers and researchers better understand biological systems, fasten research, and make educated decisions that drive innovation, optimize processes, and improve healthcare outcomes. Bioengineering professionals can uncover hidden patterns, predict outcomes, simulate scenarios, and tailor interventions in previously unattainable ways by harnessing the computational power of AI, leading to significant advances in both theory and practical applications within the field. Let us discuss various aspects:

- **Data Collection and Preprocessing:** Bioengineers collect data from a variety of sources, including genomes (set of genetic information), proteomics, clinical records, and imaging data. This information could be big, complex, and multidimensional. It is pre-processed before analysis to ensure consistency, accuracy, and compatibility across multiple data formats.
- **Feature Extraction:** Data in bioengineering frequently contains a large number of variables or features (generally embedded in the chemical structure). AI systems can extract relevant information from raw data automatically. In genomics, for example, AI can uncover relevant genetic markers or sequences linked to specific traits or disorders.
- **Machine Learning Algorithms:** Machine learning algorithms powered by AI are applied to pre-processed data. These algorithms are intended to recognize and learn patterns, correlations, and trends in data.
  AI involves several method domains, such as reasoning, knowledge representation, solution search, and, among them, a fundamental paradigm of machine learning (ML). ML uses algorithms that can recognize patterns within a set of data that has been further classified. A subfield of the ML is deep learning (DL), which engages artificial neural networks (ANNs). These comprise a set of interconnected sophisticated computing elements involving "perceptons" analogous to human biological neurons, mimicking the transmission of electrical impulses in the human brain. ANNs constitute a set of nodes, each receiving a separate input, ultimately converting them to output, either singly or multi-linked using algorithms to solve problems. ANNs involve various types, including multilayer perceptron (MLP) networks, recurrent neural networks (RNNs), and convolutional neural networks (CNNs), which utilize either supervised or unsupervised training procedures [3].
- **AI in Drug Discovery:** Drug discovery involves large number of molecules and its practically very difficult to analyse such a huge variety of molecules. However, AI with help of neural networks and machine learning algorithms can simply this task with it own limitations.
  AI can recognize hit and lead compounds, and provide a quicker validation of the drug target and optimization of the drug structure design. Different applications of AI in drug discovery are depicted in Figure 10.5 [37].

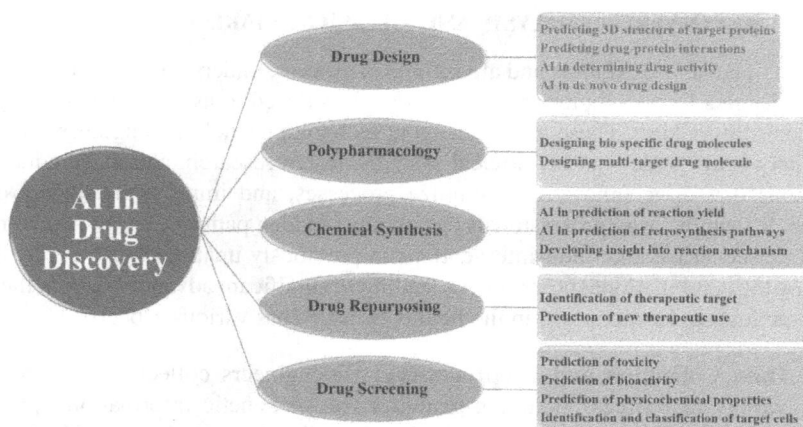

Figure 10.5 Use of AI can lead to significant advances when it comes to Drug designing, repurposing, screening, chemical synthesis, and polypharmacology.

- **Clinical Decision Support Systems (CDSS):** These are programmed with rule-based systems, fuzzy logic, artificial neural networks, Bayesian networks, as well as general machine-learning techniques [38]. CDSS with learning algorithms are currently under development to assist clinicians with their decision-making based on prior successful diagnoses, treatment, and prognostication. This domain focuses on the ethical and governance challenges arising from the development and implementation of AI-assisted CDSS within clinical- or practice-based contexts and research. Generally speaking, research encompasses knowledge-generating activities while clinical practice focuses primarily on patient care. Even though similarities exist in the ethical norms that govern both activities (e.g., obtaining informed consent and maintaining confidentiality), there are also differences in how risks and benefits are weighed up [39].

Following are some AI based software's used in this field:

- **DeepTox:** Software that predicts the toxicity of total of 12,000 drugs.
- **DeepNeuralNetQSAR:** Python-based system driven by computational tools that aid detection of the molecular activity of compounds
- **ORGANIC:** A molecular generation tool that helps to create molecules with desired properties.
- **PotentialNet:** Uses NNs to predict binding affinity of ligands.
- **Hit Dexter:** ML technique to predict molecules that might respond to biochemical assays .
- **DeltaVina:** A scoring function for rescoring drug-ligand binding affinity.
- **Neural graph fingerprint:** Helps to predict properties of novel molecules.
- **AlphaFold:** Predicts 3D structures of proteins.

10.4.3   COMPLEMENTARY ROLES OF IOT AND AI IN ENHANCING NATURAL MATERIAL APPLICATIONS

After discussing both IoT and AI now we can further move on to how we can integrate both of them together advance in bio engineering.

The Venn diagram (Figure 10.6) indicates the integration of these two fields. The complementary roles of IoT (Internet of Things) and AI (Artificial Intelligence) can greatly enhance the applications of natural materials in various industries. IoT is a vast concept encompassing too many sensors, actuators, data storage and data processing capabilities interconnected by the Internet [7]. Thus, any IoT enabled device can sense its surroundings, transmit, store and process the data gathered and act accordingly. The last step of acting accordingly is entirely dependent on processing. The true smartness of an IoT service is determined by the level of processing or acting. A non-smart IoT system will have limited capability and will be unable to evolve with the data. However, a smarter IoT system will have artificial intelligence and may serve to achieve the actual goal of automation and adaptation [40]. Following are some examples that integrate IoT with AI:

- **Sophia:** Sophia a social humanoid robot from Hanson Robotics, is very human-like and can express emotions through more than 50 facial expressions. It is capable of maintaining eye contact with the person while chatting. Sophia is the world's first robot to get a country's citizenship (Figure 10.7).
- **Moley Robotics' Robotic Kitchen:** It is an advanced fully functional robot fitted into a kitchen. It has robotic arms, an oven, a burner, a touchscreen

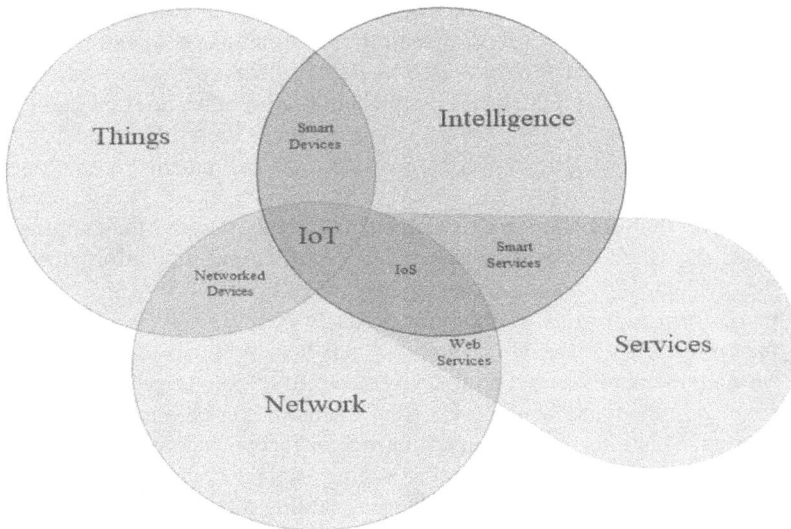

Figure 10.6   Integration of IoT and AI.

Figure 10.7 Sophia the robot: A symbol of AI's incredible advances in shaping the future.

device for human interaction, and it can cook expert-quality meals from its recipe database. Natural language processing, computer vision, shape recognition, object recognition, detection and tracking, block-chain technology to analyze inputs and responses, facial recognition, voice recognition, speech-to-text technology, obstacle recognition, haptics, and other technologies are widely used in these robots to enable them to function effectively.

- **Agriculture and fertilizer management:** IoT sensors can detect soil calcium and phosphate levels and send the information to a central system. This data can be used to optimize the use of calcium phosphate-based fertilizers. AI can use IoT data, in conjunction with other relevant elements (such as weather and crop variety), to generate exact fertilizer recommendations, boosting optimal nutrient consumption and reducing environmental impact.

- **Engineering and the development of materials:** During the creation of new materials, IoT-enabled experiments can offer real-time information on the crystallization and mechanical characteristics of various calcium phosphate formulations. This data can be analysed by AI to design calcium phosphate based material for dental fillings bone gratings, etc.

- **Management of resources and sustainability:** The usage of natural resources in the manufacture and use of materials can be tracked using IoT sensors. This information can aid in more effective resource management and waste minimization. In order to reduce resource consumption, improve processes, and increase the overall sustainability of natural material applications, AI can analyse the data gathered from IoT devices.

10.5 APPLICATIONS OF IOT AND AI WITH NATURAL MATERIALS

10.5.1 SMART MONITORING OF NATURAL MATERIAL IMPLANTS

A new class of smart medical implants with revolutionary features and cutting-edge functions is urgently needed. Here, the idea of "self-aware implants" is put up to allow for the development of a new breed of implantable multifunctional metamaterial devices that can respond to their surroundings, empower themselves, and self-monitor their health. By combining mechanical metamaterial design paradigms with nano energy harvesting paradigms, these functions are made possible creating proof-of-concept interbody spinal fusion cage implants with self-sensing, self-powering, and mechanical tunability capabilities.

Bench-top testing is performed using synthetic biomimetic and human cadaver spine models to evaluate the electrical and mechanical performance of the developed patient-specific metamaterial implants. The results show that the self-aware cage implants can diagnose bone healing process using the voltage signals generated internally through their built-in contact-electrification mechanisms. The voltage and current generated by the implants under the axial compression forces of the spine models reach 9.2 V and 4.9 nA, respectively. The metamaterial implants can serve as triboelectric nanogenerators to empower low-power electronics [41].

With the rapid development of the Internet of Things (IoT), the number of sensors utilized for the IoT is expected to exceed 200 billion by 2025. Thus, sustainable energy supplies without the recharging and replacement of the charge storage device have become increasingly important. Among various energy harvesters, the triboelectric nanogenerator (TENG) has attracted considerable attention due to its high instantaneous output power, broad selection of available materials, eco-friendly and inexpensive fabrication process, and various working modes customized for target applications. The TENG harvests electrical energy from wasted mechanical energy in the ambient environment [42].

AI combined with data received from the IoT system can help monitor the efficiency of the materials used in these generators thus helping with efficient improvements and growth of IoT and AI as a whole.

There are many ways in which in which AI and IoT can be integrated with natural materials for smart monitoring:

- **Miniaturized sensors:** that can measure a variety of parameters, including temperature, pressure, strain, pH levels, and biological markers, can be included in implants. These sensors can offer ongoing information on the condition of the implant and how it interacts with the body, thus telling us about the efficiency of implants and hence that if the materials.
- **Wireless connectivity:** Implants can be equipped with wireless communication capabilities, allowing them to transmit data to external devices such as smartphones, tablets, or medical equipment. This enables remote monitoring by healthcare professionals and patients themselves.
- **Integration with Electronic Health Records (EHRs):** Smart implants can be integrated with a patient's EHR, giving a thorough overview of the

patient's health and implant performance over time. The individualized treatment regimens may benefit from this integration.

- **Energy harvesting:** To fuel the integrated sensors and communication systems, implants can use energy harvesting techniques, such as using body heat or motion. This eliminates the need for external power sources or regular battery replacements.
- **Biocompatibility and safety:** It's critical that the additional smart components do not jeopardize the natural material implant's biocompatibility and safety. To ensure that the smart features do not have negative effects or issues, extensive testing, and optimization are needed.

10.5.2   PREDICTIVE MODELING FOR NATURAL MATERIAL DEGRADATION

IoT can be one of most efficient solution to monitor and predict degradation of natural materials. This can be of great use as the stability and life period of materials plays a great role in determining their use in different applications. The second law of thermodynamics implies that any animate and inanimate systems degrade and inevitably stops functioning. It is irreversible over time that can be labeled as the degradation arrow of time. From perspective of products reliability design, it is essential to build appropriate models of describing the degradation arrow of time. The current modelling approaches mainly include the model-driven (having assumed forms based on cognitive experience of mankind) and data-driven (using data learning techniques without form hypothesis) approaches [43]. The data can first be gathered by through sensors and the be transferred via IoT to the Ai models. Based on the characteristics of the data and the degradation process, select an appropriate predictive modeling technique. Common methods include time-series analysis for degradation prediction, linear regression, decision trees, random forests, support vector machines, and neural networks. After the model is selected, Split the data into training and validation sets. Train the selected model using the training data, optimizing its parameters to minimize the prediction error or loss function. Once the model is trained and validated, you can use it to predict material degradation over time. Continuously monitor the material's exposure conditions and update the model as new data becomes available. Gain insights into the degradation process by interpreting the model's predictions and feature importance. Understand which factors contribute most to material degradation and make informed decisions accordingly. By these methods the degradation of natural materials can be quantified, and hence be used at places with desired applications.

10.6   CHALLENGES AND FUTURE DISCUSSIONS

With ever expanding possibilities and ideas, there are a lot of changes and developments one can expect from IoT and AI. The maturation of these technologies has the capacity to reshape how we engage with our surroundings, seamlessly merging the physical and digital realms. Through AI-driven automation, numerous tasks stand to benefit from accelerated and optimized execution, resulting in time savings and cost-reduction.

IoT integrated health bands are one of the most common example in our daily lives when it comes to using IoT and AI with biotech engineering applications. But one can expect AI to be integrated in such a way that allows us to create smart prosthetics, smart nutrition and digestion monitoring, assistive tissue healing, predicting and fighting deadly diseases such as HIV [44], and cancer [5]. Companies such as Neuralink Corp. are working on brain-to-computer interfaces (BCIs) and have a futuristic outlook towards revolutionizing the way we interact with technology.

Traditional electric fans are controlled by switches and regulators. Transmission of thoughts, ideas, emotions, pictures, and memories through BCIs, controlling heavy machinery with light hand gestures, curing diseases and impairments using the combination of microchips, sensors, and AI could make the world free from disabilities.

As AI and IoT continue to develop, they will become intertwined with our lives and will reach the horizon we have never reached before. This is an ever-growing field with a lot of potential to grow and would is the root to some mind blowing sci-fi technology.

But with this fascinating technology come various challenges. Elon Musk has referred to AI as "summoning the demon" and has advocated for responsible AI development to ensure its safe and beneficial use. AI is a powerful tool that has numerous uses, from the automation of home comforts to revolutionizing industries through data analysis and predictive modeling. But misuse of AI and the data generated by IoT devices is a major concern when it comes to the large-scale implementation of such technologies. The data and information processed by IoT and AI-integrated devices are vast and often private. This raises concerns regarding data privacy and security. Recent developments in quantum computing also discuss how the entire encryption protocol could be compromised with these powerful machines.

The Quantum Limits of Miniaturization: When Small Becomes Too Small- In the rapidly evolving landscape of technology, one undeniable trend has been the consistent reduction in the size of our devices. From bulky mainframes that once filled entire rooms to sleek, pocket-sized smartphones, miniaturization has been a hallmark of progress.

At the macroscopic level, the rules of classical physics seem to hold true. Objects have definite positions and velocities, and their behaviors are predictable and deterministic. However, as things become infinitesimally small, the foundations of classical physics are overshadowed by quantum mechanics. In the quantum world, particles can exist in multiple states simultaneously, possess a property called spin, and demonstrate an intrinsic uncertainty that prevents us from knowing both their position and momentum with absolute precision. This forms a barrier that prevents us from going any smaller than a certain point. The distance between cell walls could also be a factor in determining how small devices could be when implanted into a human body.

While all the technical aspects are being discussed, it is important to also address the social impact and how it could affect the lives of people both collectively and individually. This is a rapidly growing field that will integrate into our lives at a swift pace, but the concern lies in the level of acceptance the new technology will receive. As the development of projects like Neuralink progresses, they face considerable backlash from individuals who believe that brain implants are unethical and pose a threat to the future and the general quality of life. It is a reasonable fear, as the uses of AI and IoT are practically endless. People are concerned about the potential threat to their jobs, as they could be replaced by AI that works 24×7 without any salary. The idea of putting a chip inside ones brain that could have access to your mind and memory is a scary thought.

10.7 CONCLUSION

In conclusion, this chapter teaches us about two unique technologies, i.e., AI and IoT, and they may be used to our advantage when it comes to bioengineering with natural materials. We learned how IoT and AI collaborate to find solutions, much like two friends working together to solve a puzzle.

Bioengineering greatly benefits from the use of natural materials. They have a lot of applications, but there are some obstacles also that one has to face for making use of natural materials. Imagine using AI and IoT watch over things like natural material implants. It is like to having a guardian who is constantly watching over them. We can anticipate when these materials might no longer function properly, allowing us to be ready.

We have also learnt about various sensors that are being used and will be used in future for betterment of healthcare services. Still there are many things we need to figure out, but we are eager for what comes next. We can create amazing solutions for several problems we face by using IoT and AI. In a nutshell, this chapter tells us that how we can bring AI and IoT together to make bioengineering even more fantastic.

REFERENCES

1. F. Yaman, A. Adler, and J. Beal. Opportunities and challenges in applying artificial intelligence to bioengineering. *Automated Reasoning for Systems Biology and Medicine*, 425–452, 2019.
2. C. Janiesch, P. Zschech, and K. Heinrich. Machine learning and deep learning. *Electronic Markets*, **31**(3):685–695, 2021.
3. D. Paul, G. Sanap, S. Shenoy, D. Kalyane, K. Kalia, and R. K. Tekade. Artificial intelligence in drug discovery and development. *Drug Discovery Today*, **26**(1):80–93, 2021.
4. A. K. Philip and M. Faiyazuddin. *An overview of artificial intelligence in drug development*. 1–8, 2023.
5. Y. Kumar, A. Koul, R. Singla, and M. F. Ijaz. Artificial intelligence in disease diagnosis: A systematic literature review, synthesizing framework and future research agenda. *Journal of Ambient Intelligence and Humanized Computing*, **14**(7):8459–8486, 2023.

6. D. A. Hashimoto, G. Rosman, D. Rus, and O. R. Meireles. Artificial intelligence in surgery: Promises and perils. *Annals Surgery*, **268**(1):70–76, 2018.

7. P. Sethi and S. R. Sarangi. Internet of things: Architectures, protocols, and applications. *Journal of Electrical and Computer Engineering*, **2017**:1–25, 2017.

8. R. A. Alzafarani and G. A. Alyahya. Energy efficient IoT home monitoring and automation system. In 2018 15th Learning and Technology Conference (L&T). *IEEE*, 107–111, 2018.

9. S. Kumar, P. Tiwari, and M. Zymbler. Internet of things is a revolutionary approach for future technology enhancement: A review. *Journal of Big Data*, **6**(1), 2019.

10. J. Srivastava, S. Routray, S. Ahmad, and Md. M. Waris. Internet of medical things (IoMT)-based smart healthcare system: Trends and progress. *Computational Intelligence and Neuroscience*, **2022**:1–17, 2022.

11. S. A. Wagan, J. Koo, I. F. Siddiqui, M. Attique, D. R. Shin, and N. M. F. Qureshi. Internet of medical things and trending converged technologies: A comprehensive review on real-time applications. *Journal of King Saud University, Computer and Information Sciences*, **34**(10):9228–9251, 2022.

12. R. Dwivedi, D. Mehrotra, and S. Chandra. Potential of internet of medical things (IoMT) applications in building a smart healthcare system: A systematic review. *Journal of Oral Biology and Craniofacial Research*, **12**(2):302–318, 2022.

13. A. Tyagi, S. Semwal, and G. Shah. A Review of EEG Sensors Used for Data Acquisition. *Journal of Computer Applications (IJCA)*, 13–17, 08 2012.

14. F. Laport, A. Dapena, P. M. Castro, F. J. Vazquez-Araujo, and D. Iglesia. A prototype of EEG system for IoT. *International Journal of Neural Systems*, **30**(07):2050018, 2020.

15. A. Kumar and G. Jangir. *EEG Signal Based System to Control Home Appliances*. 2454–9150, 2018.

16. S. He, L. G. Leanse, and Y. Feng. Artificial intelligence and machine learning assisted drug delivery for effective treatment of infectious diseases. *Advanced Drug Delivery Reviews*, **178**:113922, 2021.

17. A. A. Arabi. Artificial intelligence in drug design: Algorithms, applications, challenges and ethics. *Future Drug Discovery*, FDD59, **3**(2), 2021.

18. Y. C. Yang, S. Ul Islam, A. Noor, S. Khan, W. Afsar, and S. Nazir. Influential usage of big data and artificial intelligence in healthcare. *Computational and Mathematical Methods in Medicine*, **2021**:1–13, 2021.

19. J. Tang, A. LeBel, S. Jain, and A. G. Huth. Semantic reconstruction of continuous language from non-invasive brain recordings. *Nature Neuroscience*, **26**(5):858–866, 2023.

20. B. Fiani, T. Reardon, B. Ayres, D. Cline, and S. R. Sitto. An examination of prospective uses and future directions of neuralink: The brain-machine interface. *Cureus*, **13**(3), March 2021.

21. Play Studio. Neuralink — neuralink.com. https://neuralink.com/. [Accessed 15-08-2023].

22. T. Shaik, X. Tao, N. Higgins, L. Li, R. Gururajan, X. Zhou, and U. R. Acharya. Remote patient monitoring using artificial intelligence: Current state, applications, and challenges. *WIREs Data Mining and Knowledge Discovery*, **13**(2):e1485, 2023.

23. Y. Kumar, A. Koul, R. Singla, and M. F. Ijaz. Artificial intelligence in disease diagnosis: A systematic literature review, synthesizing framework and future research agenda. *Journal of Ambient Intelligence and Humanized Computing*, **14**(7):8459–8486, 2022.

24. J. Jeong, J. H. Kim, J. H. Shim, N. S. Hwang, and C. Y. Heo. Bioactive calcium phosphate materials and applications in bone egeneration. *Biomaterials Research*, **23**(1):4, 2019.

25. C. Mirela G. Nobre, N. Pütz, and M. Hannig. Adhesion of hydroxyapatite nanoparticles to dental materials under oral conditions. *Scanning*, **2020**:1–12, 2020.

26. M. P. Wallen, S. R. Gomersall, S. E. Keating, U. Wisloff, and J. S. Coombes. Accuracy of heart rate watches: Implications for weight management. *PLOS ONE*, **11**(5):e0154420, 2016.

27. D. R. Bassett, L. P. Toth, S. R. LaMunion, and S. E. Crouter. Step counting: A review of measurement considerations and health-related applications. *Sports Medicine*, **47**(7):1303–1315, 2016.

28. P. Casale, O. Pujol, and P. Radeva. Human activity recognition from accelerometer data using a wearable device. In J. Vitrià, J. M. Sanches, and M. Hernández, editors, *Pattern Recognition and Image Analysis*, 289–296, Berlin, Heidelberg, Springer Berlin Heidelberg, 2011.

29. M. Serhani, H. El Kassabi, H. Ismail, and R. Nujum. Ecg monitoring systems: Review, architecture, processes, and key challenges. *Sensors*, **20**:1796, 2020.

30. C.-Y. Huang, M.-C. Chan, C.-Y. Chen, and B.-S. Lin. Novel wearable and wireless ring-type pulse oximeter with multi-detectors. *Sensors*, **14**(9):17586–17599, 2014.

31. Nguyen Canh Minh, To Hieu Dao, Duc Nghia Tran, Nguyen Quang Huy, Nguyen Thi Thu, and Duc Tan Tran. Evaluation of smartphone and smartwatch accelerometer data in activity classification. In *2021 8th NAFOSTED Conference on Information and Computer Science (NICS)*, pages 33–38, 2021.

32. A. Umek and A. Kos. Validation of smartphone gyroscopes for mobile biofeedback applications. *Personal and Ubiquitous Computing*, 20:657–666, 10 2016.

33. I. Ahmed, N. Jiang, X. Shao, M. Elsherif, F. Alam, A. Salih, H. Butt, and A. Yetisen. Recent advances in optical sensors for continuous glucose monitoring. *Sensors and Diagnostics*, 1, 08 2022.

34. M. Raez, Md. Hussain, and F. Mohd-Yasin. Techniques of emg signal analysis: Detection, processing, classification and applications. *Biological Procedures Online*, **8**:11–35, 2006.

35. Y. Yoonsik, S. Chae, J. Shim, and T.-D. Han. *Emg Sensor-Based Two-Hand Smart Watch Interaction*. 73–74, 2015.

36. G. Coulby, A. Clear, O. Jones, F. Young, S. Stuart, and A. Godfrey. Towards remote healthcare monitoring using accessible IoT technology: state-of-the-art, insights and experimental design. *BioMedical Engineering OnLine*, **19**(1):80, 2020.

37. K.-K. Mak and M. R. Pichika. Artificial intelligence in drug development: Present status and future prospects. *Drug Discovery Today*, **24**(3):773–780, 2019.

38. K. B. Wagholikar, V. Sundararajan, and A. W. Deshpande. Modeling paradigms for medical diagnostic decision support: A survey and future directions. *Journal of Medical Systems*, **36**(5):3029–3049, 2011.

39. T. Lysaght, H. Y. Lim, V. Xafis, and K. Y. Ngiam. AI-assisted decisionmaking in healthcare. *Asian Bioethics Review*, **11**(3):299–314, 2019.

40. A. Ghosh, D. Chakraborty, and A. Law. Artificial intelligence in internet of things. *CAAI Transactions on Intelligence Technology*, **3**(4):208–218, 2018.

41. K. Barri, Q. Zhang, I. Swink, Y. Aucie, K. Holmberg, R. Sauber, D. T. Altman, B. C. Cheng, Z. L. Wang, and A. H. Alavi. Patient-specific self-powered metamaterial

implants for detecting bone healing progress. *Advanced Functional Materials*, **32**(32): 2203533, 2022.

42. W.-G. Kim, D.-W. Kim, I.-W. Tcho, J.-K. Kim, M.-S. Kim, and Y.-K. Choi. Triboelectric nanogenerator: Structure, mechanism, and applications. *ACS Nano*, **15**(1):258–287, 2021.

43. R. Kang, W. Gong, and Y. Chen. Model-driven degradation modeling approaches: Investigation and review. *Chinese Journal of Aeronautics*, **33**(4):1137–1153, 2020.

44. F. Yaman, A. Adler, and J. Beal. Opportunities and Challenges in Applying Artificial Intelligence to Bioengineering, 425–452. 2019.

Index

For Product Safety Concerns and Information please contact our EU
representative GPSR@taylorandfrancis.com
Taylor & Francis Verlag GmbH, Kaufingerstraße 24, 80331 München, Germany